ROUTLEDGE LIBRARY EDITIONS: INTERNATIONAL SECURITY STUDIES

Volume 9

THE FUTURE OF SEA POWER

THE FUTURE OF SEA POWER

ERIC GROVE

LONDON AND NEW YORK

First published in 1990 by Routledge

This edition first published in 2021
by Routledge
2 Park Square, Milton Park, Abingdon, Oxon OX14 4RN

and by Routledge
52 Vanderbilt Avenue, New York, NY 10017

Routledge is an imprint of the Taylor & Francis Group, an informa business

© 1990 Eric Grove

All rights reserved. No part of this book may be reprinted or reproduced or utilised in any form or by any electronic, mechanical, or other means, now known or hereafter invented, including photocopying and recording, or in any information storage or retrieval system, without permission in writing from the publishers.

Trademark notice: Product or corporate names may be trademarks or registered trademarks, and are used only for identification and explanation without intent to infringe.

British Library Cataloguing in Publication Data
A catalogue record for this book is available from the British Library

ISBN: 978-0-367-68499-0 (Set)
ISBN: 978-1-00-316169-1 (Set) (ebk)
ISBN: 978-0-367-71116-0 (Volume 9) (hbk)
ISBN: 978-0-367-71118-4 (Volume 9) (pbk)
ISBN: 978-1-00-314938-5 (Volume 9) (ebk)

Publisher's Note
The publisher has gone to great lengths to ensure the quality of this reprint but points out that some imperfections in the original copies may be apparent.

Disclaimer
The publisher has made every effort to trace copyright holders and would welcome correspondence from those they have been unable to trace.

The Future of Sea Power

Eric Grove

Routledge
London

First published in 1990
by Routledge
11 New Fetter Lane, London EC4P 4EE

© 1990 Eric Grove

Printed in Great Britain by
Richard Clay Ltd, Bungay, Suffolk

All rights reserved. No part of this book may be reprinted or reproduced or utilized in any form or by any electronic, mechanical, or other means, now known or hereafter invented, including photocopying and recording, or in any information storage or retrieval system, without permission in writing from the publishers.

British Library Cataloguing in Publication Data

Grove, Eric
  The future of sea power.
  1. Sea – power
  I. Title
  359
  ISBN 0-415-00482-9

To Elizabeth

Contents

List of tables and figures	ix
Foreword by Admiral Sir Julian Oswald	xi
Acknowledgements	xiii
Part I: The state of the art	1
1 Sea power in the modern world	3
2 Maritime strategy	11
Part II: The uses of the sea	29
3 Economic uses of the sea	31
4 The military medium	46
Part III: The changing shape of naval war	59
5 Weapon systems	61
6 Platforms	96
Part IV: The evolving environment	151
7 The international and legal context	153
8 Economic stimuli and constraints	172

Contents

Part V: Navies in peace and war   185

 9 Navies in peace   187

10 Navies in war   199

Part VI: Conclusion   219

11 The future of sea power – three theoretical frameworks   221

Notes and references   242

Acronyms and abbreviations   268

Index   272

Tables and figures

Table

6.1 Most promising future seaborne platform options   122

Figures

3.1 Main oil movements by sea, 1987   32
11.1 The 'use of the sea' triangle I   234
11.2 The 'use of the sea' triangle II   236

Foreword

by Admiral Sir Julian Oswald, GCB, ADC,
Chief of Naval Staff and First Sea Lord

It is, I suppose, axiomatic that in relation to Sea Power, its place in the world and its actual or potential effect on the lives of people, both Maritime Strategy and the Structure of Navies will be constantly under review. Thus a book which combines careful analysis of the structure and components of sea power and a critical, questioning look at the strategy which governs navies' possible employments will always be important. It is particularly to be welcomed on two further scores. First it comes from the pen of Eric Grove, an experienced commentator, distinguished strategist, and most painstaking author who revels in this particular subject, and whose background is ideal. Additionally, it comes to us at a time when the whole structure of inter-state, inter-allied, and inter-bloc politico-military relationships is undergoing traumatic change. It is therefore essential that we should look ahead and ask ourselves – 'What might we expect of navies in the 21st Century?' For the United Kingdom, long term exponent of maritime power and with a unique geographical position and sea heritage, the question is of the utmost importance. Ships on the drawing board today may well be in the Fleet in 2030. Will they be adaptable to the requirements of that age? What might those requirements be? Other countries too must answer difficult questions if they are to invest sensibly in sea power.

Arms control initiatives now showering down upon us may well start by altering the balance of forces (and both the likelihood and nature of conflict) on land – but they will soon introduce seaborne considerations as well. Some measures may seek to reduce weapon systems at sea, others may encourage the deployment of weapons on maritime platforms to compensate for those removed from land. In any event it seems likely that the sea will remain host to all or part of the strategic deterrents of both the superpowers, France, the United Kingdom and perhaps others. This of itself will greatly influence maritime operations.

Asymmetric cuts in conventional forces in Europe may reduce the

Foreword

risk of attack but, paradoxically, increase the importance of reinforcement across the Atlantic and Channel. Retaining necessary and credible naval forces to effect this will necessitate not only presenting a convincing and coherent Alliance Maritime Strategy and rationale to our political leaders, but also insisting, in discussion with the Soviet bloc, that the entirely different nature of NATO and the Warsaw Pact – the one on maritime Alliance, the other a great continental power block – be recognized.

Not to be forgotten is the impact of navies outside the NATO/Warsaw Pact area. Can we judge that as a better relationship and understanding grows between the superpowers and their allies, they will turn their efforts more to the prevention of war and the resolution of conflict in the rest of the world? If they do they will find very many navies, some far from small and many well-equipped and capable, out there. What interactions can be expect, and how will these perceptions effect naval development around the world?

The questions are endless. Some are unanswerable. To others we can give a qualified reply. But if we are to get even that right we must be absolutely clear about which principles of Maritime Structure and Strategy will endure, and which may not. To guide us is the fascinating and demanding task which Mr Grove has set himself.

Julian Oswald

Acknowledgements

Three good friends played crucial roles in this book. Philip Towle of Queen's College, Cambridge first suggested I should write it. Evan Davies of the Britannia Royal Naval College, Dartmouth, who was to have been co-author, was of great assistance in defining its basic structure. Geoffrey Till of the Royal Naval College, Greenwich provided useful suggestions for improving the final draft. Michael Ranken and R. S. Woolrych of the British Maritime Charitable Foundation were more than helpful with information and advice on merchant shipping matters. British Petroleum kindly allowed reproduction of their chart of oil movements by sea and D. W. Waters and Ken Booth both very graciously assented to extended quotations of their work. Norman Friedman, a name to be found often in the footnotes, has always been a constant source of insight both in his written works and our private discussions. Peter Sowden of Croom Helm, now Routledge, tolerated with amazingly good grace the very long, but unavoidable, delays in final production of the manuscript. My mother and father typed, re-typed, and printed out the various drafts in their usual efficient manner. My wife, to whom this book is dedicated, put up with the burden of being an obsessive author's partner. To all the above very many thanks. No one but the author is responsible for the facts or judgements that follow: they do not reflect the policy of any organization or institution.

Part I
The state of the art

Chapter 1

Sea power in the modern world

'Sea Power' means different things to different people. It can be an almost mystical concept, a magic formula to be mouthed in awe-struck tones to scare away evil spirits such as defence ministers with non-naval priorities or air force officers with alternative means on offer of providing a state's military power on or across the oceans. To others, following in the well-trodden footsteps of Alfred Thayer Mahan, 'sea power' represents a more coherent but equally universal concept, an interlocking system of forms of sea use, military and civil, which has a unique contribution to making powers great or greater still. This book will take a narrower and more measured approach. In series with its two earlier companions on 'air power' and 'land warfare' it will treat 'sea power' as a military concept, that form of military power that is deployed at or from the sea. As Mahan himself put it, 'the history of sea power, while embracing in its broad sweep all that tends to make a people great upon the sea or by the sea, is largely a military history';[1] equally, the future emphasized here will be the military future of sea power, although the relationship of naval power with the various forms of sea use must also be considered.

The whole subject has become rather more complex than it once seemed in the age of Mahan. For a number of economic reasons, which are dealt with more fully in Chapter 3, the states that possess the world's most powerful navies do not possess the world's largest merchant fleets. Despite many state sponsored attempts to prove Mahan correct in his famous assertion that, 'The necessity of a navy in the restricted sense of the word, springs, therefore, from the existence of a peaceful shipping and disappears with it',[2] the world's most powerful naval power was outranked at the end of 1986 in merchant shipping by Liberia, Panama, Japan, and Greece.[3] Liberia, with 97.8 million deadweight tons of ships and 15.6 per cent of the world's total tonnage, deployed a military fleet of six patrol craft, none bigger than 50 tons, and a land-based helicopter.

The state of the art

The largest weapon available was an 81 mm mortar. Panama, with 69 million tons of merchantmen, 11 per cent of the world's tonnage, has eight small patrol boats and eight transports; one of the latter is a former shrimp boat.[4] The two major mercantile marines of the world seem to prosper with nothing more than a coastguard back home. Only when we come to the world's number three, Japan, with almost 56 million deadweight tons of merchant ships and almost 9 per cent of world tonnage, do we find the possessor of a significant military navy, one that 'has tended to look more and more like an ocean going navy'.[5] Nevertheless, its operations are still severely circumscribed and it is still significantly smaller and less capable than the navies of Britain and France, countries which no longer make it into the merchant shipping top ten (unless Britain's dependencies are included). Something has clearly gone wrong with the Mahanian paradigm.

One factor has been the growing internationalization of marine activities. Mahan drew his lessons from the great days of classical mercantilism and the clashes of autarkic empires competing for slices of a finite 'cake' of maritime trade. He thought this might well recur in the twentieth century. In a way it did, but the defeat of the German and Japanese attempts to carve out continental and maritime variations on this theme led to a new world of economic neo-liberalism that still, despite numerous vicissitudes, survives. In this environment the flag a ship carries is very often not a reflection of the identity of the owner. If US-owned 'flag of convenience' vessels are taken into account, the tonnage of the American mercantile marine more than doubles to over 65 million tons; but that of Japan increases by an almost similar margin, to over 90 million. Even taking beneficial ownership into account, therefore, the United States remains only third in merchant ship ownership, behind Japan and Greece.[6]

Why then is she the number one naval power, closely followed by the USSR, with a mere 23.4 million tons of merchant ships. The fundamental fact of twentieth-century sea power is that a country's naval capability is a direct reflection of its sheer economic power in all senses and that that power inevitably reflects its control and exploitation of large land masses. Professor Paul Kennedy has ably demonstrated that Halford Mackinder not Alfred Mahan proved the surer prophet of the new century.[7] In 1902, in his *Britain and the British Seas*, Mackinder argued that Britain's sea power rested on fragile foundations in an era of 'vast Powers, broad based upon the resources of half continents . . . mere insularity gives no indefeasible title to marine sovereignty.'[8] In 1904, before the Royal Geographical Society, he went further by suggesting that 'When historians of the remote future come to look back . . ., it may well be that they will describe the last 400 years as the Columbian epoch, and will say that it ended soon after the year 1900.' Mackinder argued that with the world now virtually fully explored and divided up, 'we are for the first time in a position to attempt, with some degree of completeness, a correlation between the larger geographical and the larger historical generalisations.'[9] In other words, as Kennedy has rather more clearly put it, 'efficiency and internal development would replace expansionism as the main aim of modern states and . . . size and numbers would be more accurately reflected in the sphere of international developments.'[10] The railway already meant that for the first time in human history, transport by land was as (if not more) effective. And in some ways it was as economical as transport by water. The beginnings of air (and road motor) transport promised to confirm this secular change. In one of the most perceptive comments by a member of Mackinder's audience, Leo Amery, while insisting that a great power must possess sea power as well as land power, accepted that 'sea power alone, if it is not based on great industry and has a great population behind it, is too weak for offence to really maintain itself in the world struggle'. In future, once the transport revolution had taken place 'the successful powers will be those who have the greatest industrial base. It will not matter whether they are in the centre of a continent or on an island; those people who have the industrial power and the power of invention and of science will be able to defeat all others.'[11]

And so they have. The world's superpowers, the USA and the USSR, stand supreme upon the world's oceans, their navies confronting each other around the globe as global expressions of the fundamental politico-strategic split that divides the planet. Both sides' navies are more expressions of a power that has its roots elsewhere than, as the Royal Navy was at the height of its power, the guardians of the foundations of national economic power itself, i.e. seaborne trade. Nothing better symbolizes the foundations of sea power in the modern world than the nature of much of both sides' current naval investment. The superpowers' ballistic missile firing submarines prowl their patrol areas ready at a moment's notice to destroy the political and economic heartland of the other side. They use the hiding place provided by the sea to make themselves the least vulnerable and most stabilizing form of long-range nuclear striking power: they might even be regarded as a form of air power. For the USSR, perhaps the one form of sea use that is fundamental to its national security is the ability to use the seas to deploy this form of strategic reserve as a final sanction against all-out Western nuclear

The state of the art

attack. Certainly the USA seems to think so in its recent articulations of how to exert potentially decisive maritime pressure on the USSR in a conventional war.

Modern technology, as explained more fully in Chapter 5, has revolutionized the ability of sea-based forces to 'thrust power ashore', especially in the form of bombardment. 'Sea power' can thus directly affect the land battle in immediate and short-term ways that it could not do, except on the coastline itself, until the middle decades of this century. Before that 'sea power' tended to be an indirect factor in the campaign ashore, maintaining sea-dependent states in the war so that they could exert direct military pressure on the enemy in various ways, be it subsidies to allies, giving battle on land or, in later years, bombing from the air. Even at the height of the Columbian era in the eighteenth and early nineteenth centuries the last and greatest maritime empire, the British, had to maintain armies in the field against France if she was to ensure success: the one war where she did not do this was the one she lost – the War of American Independence.[12] Sea power prevented the wars against Napoleon being lost but Mahan's famous 'storm tossed ships upon which the Grand Army never looked'[13] could do little in themselves to win them.

When that greatest of all maritime strategists, Sir Julian Corbett, wrote up in 1911 the lectures of history and strategy with which he had been intriguing (and annoying) the officers of the Naval War College for the previous five years, he stated the situation plainly: 'Since men live upon the land and not upon the sea, great issues between nations at war have always been decided – except in the rarest cases – by what your army can do against your territory and national life or else by the fear of what the fleet makes it possible for your army to do.'[14]

The events of 1914–18 bore Corbett out. Lord Esher set out the strategic situation only ten days into the war. 'No one,' he wrote, '. . . believes that the naval forces of the allies can force a decision, or do more than render the lives of the Allied peoples secure and that of the enemy impossible. We shall remain immune from attack. The French will be free to receive supplies from oversea. The Germans will become a beleaguered garrison. These achievements are those of modern sea command.'[15] Esher doubted if a major battle would take place and when it did, off Jutland in 1916, the British fleet commander wisely did not risk losing the fruits of Britain's current strategic situation for the somewhat limited extra gains that would have been obtained from destroying the German battlefleet. Britain was thus able to continue to draw on the resources of the world to deploy the Imperial Army that, in its finest hour, defeated the full might of the German Army in the summer and autumn of 1918. When, following these land defeats, the Germans gave up most of their battlefleet to the British without a fight the Admiralty was moved to send a congratulatory signal to the Grand Fleet 'on a triumph to which history knows no parallel' which would 'remain for all time the example of the wonderful silence and sureness with which sea power attains its ends'.[16] A soldier from Amiens or the Hindenburg line might have disputed the silence if not the sureness.[17]

The Admiralty in its November message spoke of the pressure exerted by the Grand Fleet itself on Germany, and much has been made of the supposed effects of the blockade on the German war economy and the morale of her people.[18] Sadly the facts do not bear this out. Numerous historians have pointed out the limitations of blockade against a power like Germany, which for most of the war controlled most of the continent.[19] German inefficiency in food production and distribution was more responsible for shortages than the blockade itself.[20] More important still, the fact that her resources were tied up in a debilitating land war was a vital factor: 'civilian sufferings from the blockade would have been far fewer had not the great military campaigns swallowed up such astronomical amounts of foodstuffs, industrial products and especially the men needed to farm the land.'[21] Of course, if the British could not win the war at sea they could certainly lose it, and the German submarine offensive in 1917 came close to achieving that success.[22]

In the second great twentieth-century war of 1939–45 the same pattern was repeated. Britain was, with some considerable effort both in the air and at sea, able to survive and act as a forward base for the projection of US power into Europe, but it was the latter and even more so the great land battles on the Eastern Front that defeated the might of Hitler's Reich. What was survival to the British was an exercise in seaborne power projection to the Americans. In the Pacific the United States demonstrated even more clearly the new shape of sea power, as a three dimensional fleet of unparalleled power and reach covered a succession of amphibious landings that cut the Japanese Empire in two and gained the bases for an aerial assault that eventually, with nuclear firepower, proved decisive. In addition, and almost as an afterthought, given Japan's extraordinary power of ignoring strategic realities and coming close to committing national suicide, in the most successful war against seaborne trade in history, Allied submarines sank over 1150 ships of almost 5 million tons. Aircraft sank almost 750 more of almost 2.5 million tons.[23] By the end of the War intensive Allied mining had bottled up the remaining ships and a maritime empire's arteries had ceased to function. Japan had been effectively strangled. How long she

The state of the art

could have continued fighting without the *coups de grace* of Hiroshima and Nagasaki is an interesting point.

The major lessons of the two world wars of the twentieth century seem to be that sea power, while a necessary part of a great power's armoury if it wishes to project military power around the globe, is far from sufficient on its own to assure victory in war. Indeed, overdependence on the sea far from being an asset was a positive vulnerability only limited by extraordinary national determination and the suicidal propensities of national culture. If extra pressure is required in these circumstances perhaps only nuclear weapons are capable of providing it.

The year 1945, therefore, in the end turned out to be a victory for air power, albeit air power that could only operate through seaborne power projection. The advent of nuclear weapons confirmed the challenge of the salience of sea power. As the kiloton range nuclear bombs became megaton range thermonuclear devices in the 1950s, serious doubts arose in the minds of many as to whether a future war would last long enough for the sea to be relevant at all. Sea power was slow acting and the next war might be over in hours. For a time navies argued that 'broken backed' war might still be possible and necessary after the initial nuclear strikes as both sides licked their dreadful wounds and prepared for a second more intensive phase, but this carried less and less conviction and proved a weak foundation for the continued survival of naval power. Better to climb onto the nuclear bandwagon and substitute sea-based systems for the forward air bases. This was successfully achieved in the 1950s and 1960s by the navies of both superpowers.[24] The technological emphasis varied but the strategic logic remained the same: sea power as a form of air power against Mackinder's heartland. It was not surprising that the relationship of superpower navies with sister services, be they the USAF or the Soviet Strategic Rocket Forces, ranged from uneasy to downright bad.

Naval officers, however, were the first to see a fundamental truth of the nuclear age. Nuclear war is so clearly self-destructive that it can only be credibly invoked as a threat when the most vital national interests are threatened, if indeed then. Perhaps nuclear weapons are only fully credible as necessary deterrents to nuclear attack. If states wish, or need, to use force to defend their interests and politico-economic health they have to have a range of capabilities much more discriminating in their effects than the instruments of nuclear Armageddon. This has a relevance both in superpower confrontation and in less equal conflicts. Whatever the context, maritime forces are often useful and sometimes indispensable for exercises of limited force. This might not ease relations much with the air force, especially if the latter sees limited war operations as its major role (cf. the controversy over British carriers in the 1960s), but it does provide one solid rationale for the navies of those states that aspire to intervene militarily beyond their boundaries.[25]

'Seaborne power projection' has thus become the main role for naval forces in the nuclear age, but the seas themselves remain crucial factors in the world economy and hence in the overall security interests of states. Although, as we shall see in Chapter 8, the level of sea *dependence* of many countries is arguable there can be little gainsaying the fact that many states have vital *interests* connected with an ability to use the sea in various ways. Moreover, interests in sea use go far wider than a few especially dependent states. The sea provides the transport medium for the vast bulk of the world's international trade, about 80 per cent by volume.[26] Over 99.5 per cent of the weight of cargoes carried over the oceans goes by sea, less than 0.5 per cent by air.[27] The free flow of these supplies is of great interest to the whole world, especially the OECD nations, whose trade tends to be more maritime than that of the communist nations of the CMEA. The USA, now rather more reliant than before on foreign sources of both manufactured goods and raw materials, sees it as being in her interest to provide protection when required to the West's seaborne trade, to the extent of lending its flag to ships under threat as in the Gulf in 1987. This 'return to the flag' and restoration of Mahanian proprieties is one of the most interesting developments of recent years and may presage a growing practice when danger threatens. Other countries, especially those like Britain whose dependencies can offer the national flag on virtual open registry (i.e. flag of convenience) terms and which can also offer naval protection from the metropolitan power, may also be in sometimes embarrassing demand by ship owners and operators.

The sea has also increased in importance as a source of useful resources. Food has always been obtained from the sea but now energy and, to a growing extent, other minerals come from the ocean and its floor. This has given the ownership of the coasts and islands of the world a new importance as these now give sole exploitation rights to huge areas of sea. Although the precise legal regime for the world's oceans remains in dispute the enclosure of large tracts of sea and ocean has already taken place. Countries need the means to assert their rights and carry out their duties in these newly-defined territorial seas and Exclusive Economic Zones, and this is perhaps the major reason for the proliferation of what might be called the institutions of sea power – the world's navies.

These factors, and the sheer growth in numbers of sovereign

states, have contributed to making navies increasingly popular. Never have there been more then there are today. Ninety years ago the first edition of *Jane's Fighting Ships* listed the vessels of a mere twenty-two different navies.[28] Ten years later, when what with Edwardian arrogance were dismissed as 'unimportant' navies (Cambodia, Costa Rica, Persia, Zanzibar, etc.) were added, the number doubled to forty-four.[29] The nineteeth issue of *Jane's* listed over 150 nations with navies and/or coastguards.[30] Sea power, as a military activity at least, is obviously a growth industry.

It is even being thought about more. One of the curious features of the growth of strategic studies over the last quarter century or so has been the relative inattention paid to sea power. Only with the publication of the US Navy's controversial 'Maritime Strategy' have we seen the growth of a strategic debate worthy of the name.[31] What is rather extraordinary about this debate is that there is so little new in the Maritime Strategy, other than the fact of its publication and the coherence and forcefulness of its drafting. As various authors have shown[32] the perceived imperatives of technology and resource constraints forced Western navies to emphasize the 'attack at source' on Soviet naval assets from the earliest days of NATO. Forward carrier and submarine operations have been a central and persistent feature of the naval strategy of the Anglo-American maritime coalition from the start. That their re-emphasis should catch so many by apparent surprise confirmed the author's assessment that there was a need to examine the future of sea power both in terms of a comprehensive assessment of the present reality and some understanding of the experience of the past.

This book will attempt to answer two main questions. How is sea power going to develop in the next few decades? Is it possible to develop a new theory of sea power for the future or are the old ideas still relevant, despite all the technological changes of the last century? Before going tentatively into the future however, like Mahan and Corbett, we must look more systematically and in some detail at the lessons of experience, so we know from where we begin.

Chapter 2

Maritime strategy

If sea power is to be used purposively then it must be through the application of maritime strategy. This can be defined as the art of directing maritime assets (i.e. those that operate on, over, or under the sea) to achieve the required political objectives. In 1911 Sir Julian Corbett set down 'some principles of maritime strategy' that were based on an intelligent reading of naval history up to that date.[1] Corbett was criticized at the time on two main grounds. First, it was said that as a civilian he had no *locus standi* to comment on such arcane arts as maritime strategy. The founder of *Jane's Fighting Ships* dealt with such criticisms devastatingly: 'it would be a bad day ... if the principle ever gets established that unless a man is an actor he is incapable of criticising the actions of a drama ... the contention would work out that "you cannot tell whether an egg is good or bad unless you are a hen".'[2]

The second criticism was one that is very familiar today: that technological progress had so fundamentally altered everything that historical parallels were inapplicable to contemporary naval affairs. If accepted, this line of argument would make naval affairs something separate from the rest of human activity, all of which is based to varying extents on experience. Progress in any field, even a fundamental breakthrough in science and technology, is usually based on an intelligent and informed reassessment of existing phenomena. If any constructive thought is to take place it must have an existing frame of reference. As Corbett himself put it: 'Having determined the normal we are in a stronger position. Any proposal can be compared with it and we can proceed to discuss clearly the weight of the factors which prompt us to depart from the normal.'[3]

The tendency to argue that 'everything has changed' is a very dangerous one. Even Corbett fell into the trap in his discussion of the defence of trade in *Some Principles*. Only six years after the publication of his book, which fell in with the current fashion in denying the continued relevance of convoy, Britain was almost defeated because of the steadfast refusal of her naval leaders to introduce a convoy system for the defence of trade. Knowledge of convoy's crucial role in 1917–18 (and 1939–45) does not mean necessarily that convoy is the only answer in all circumstances to the problem of the defence of merchant ships, but it does mean that we should be very careful indeed about departing from convoy as the core of any strategy for sea use.

The use of the sea for movement is itself the core of maritime strategy in its traditional sense. Corbett put it in slightly more general terms: 'maritime communications whether for commercial or military purposes'.[4] Control of these communications in terms of being able to achieve one's own movement and deny it to an opponent is the essence of 'command of the sea' and, as Corbett flatly stated, 'the object of naval warfare must always be directly or indirectly either to secure the command of the sea or to prevent the enemy from securing it'.[5]

There is something rather old-fashioned about the term 'command of the sea'. Some modern writers have argued that it implies a degree of dominance that modern technology makes impossible. One of the most notable of these was Admiral Stansfield Turner, perhaps the leading figure in American naval thought in the 1970s. When running the US Naval War College in 1974 he popularized the notion of 'sea control' as 'a deliberate attempt to acknowledge the limitations on ocean control brought about by the development of the submarine and the airplane'.[6] He went on to justify his new term as follows:

> The new term 'Sea Control' is intended to connote more realistic control in limited area and for limited periods of time. It is conceivable today to temporarily exert air, submarine and surface control in an area while moving ships into position to project power ashore or to re-supply overseas forces. It is no longer conceivable, except in the most limited sense, to totally control the seas for one's own use or to totally deny them to an enemy.[7]

Turner's thinking reflected both the experience of World War II and the perceived weakness of the US Navy of the post Vietnam period against the growth of Soviet naval capability of the Gorshkov era. 'Sea Control', however, is a useful concept in a more general sense in that it demonstrates the fundamental change in the capacity of states to attack shipping that was brought about by the advent of submarine and aircraft. Nevertheless, it should not lead to the complete abolition of the older term. Analysts such as Corbett clearly recognized that 'the most common situation in a naval war is that neither side has the command; . . . the normal position is not

Maritime strategy

a commanded sea but an uncommanded sea'.[8] Command of the sea is clearly a relative not an absolute term. Dr Geoffrey Till admirably encapsulates this dimension in his definition, which probably remains the best:

> Being 'in command of the sea' simply means that a navy in that happy position can exert more control over the use of the sea than can any other. The degree of command varies greatly and is primarily illustrated by the extent to which it confers the capacity to use the sea for one's own purposes and prevent the enemy using it for his.[9]

In this sense modern naval operations are still about command of the sea.

In order to secure command of the sea it is necessary to neutralize the enemy's main naval forces. The classic way of achieving this is to defeat and destroy them in battle. Mahan and many other naval historians have been fascinated by naval battles and have tended to overrate their importance. Perhaps the classic example of such misassessment is the most famous naval battle of all, Trafalgar. Nelson's great victory of 21 October 1805, when his numerically inferior fleet smashed the combined battlefleet of France and Spain, has justly been remembered as a classic of its type, although one that relied on the gross qualitative inferiority of the French and Spanish that it was Nelson's genius to observe and act upon. What is quite fascinating, however, is the way that, at the regular celebrations of the event in the modern Royal Navy, the victory is provided with an almost completely spurious strategic rationale. Senior officers solemnly lecture their audiences about how Nelson's victory saved Britain from invasion, forgetting that at the moment of victory the French Army, its invasion plans long abandoned, was already on the Danube confronting Britain's continental allies, Austria and Russia.[10]

Even Mahan clearly recognized this and Corbett found the Trafalgar campaign of such interest precisely because it was the sophisticated combination of manoeuvre and blockade that had worsted Napoleon's invasion attempt.[11] The battle achieved something, the blunting of a move on Naples, the demonstration of British naval superiority, Nelson's long desired self-immolation as an immortal national hero, but it did *not* save Britain from invasion.

Battles are means to an end not ends in themselves. Nevertheless, some battles have had momentous results. In World War II, Midway in 1942 marked the end of Japanese expansion and the demise, for a time, of her main instrument of naval warfare, her carrier striking force. The destruction of the latter was completed at the Philippine

The state of the art

Sea two years later.[12] But such decisive engagements are relatively rare. Most naval battles are but episodes of larger campaigns and there are serious difficulties in bringing even these about, especially in the modern world with its developed surveillance techniques. In 1941, before Cape Matapan, Admiral Cunningham, the British C-in-C Mediterranean Fleet, was forced into a cunning deception plan in order to prevent the Italians thinking he was about to emerge from harbour. This unwelcome news would have discouraged the enemy sortie Cunningham needed to get to grips with his opponent. Nelson had used similar deceptions with similar purposes.[13]

The development of the aircraft and the aircraft carrier made such stratagems less necessary. The carrier gave a new meaning to the traditional maxim of 'seeking out and destroying' the enemy's fleet. At Taranto in 1940 and Pearl Harbor in 1941 carrier aircraft struck at enemy fleets in harbour with strategically devastating effect. In 1945 the Japanese Navy ceased to exist in the repeated US carrier air strikes on its bases. After World War II, as it seemed to become more and more difficult to counter enemy naval assets actually at sea, and as nuclear weapons greatly increased the power of the offence, the 'attack at source' began to dominate Western naval thinking.[14] Now the new Maritime Strategy postulates – as the ultimate but not necessarily most likely threat to Soviet naval power – a conventional battle in which Western naval forces so threaten Soviet assets that the Soviet Navy is both drawn out to fight and destroyed in its bases.[15]

This is heady stuff and traditional US Navy rhetoric about 'going in harms way' is deployed to justify it. Some modern analysts find this rhetorical approach as disturbing today as Corbett found the equivalent slogans in the early years of the century. Commenting on the oft-quoted British maxim that 'the enemy's coast is our true frontier' he said, 'you might as well try to plan a campaign by singing "Rule Britannia".'[16] Corbett, in his notes for the naval officers to whom he lectured in 1906, set down the following important warning against going on the offensive for its own sake.

> In applying the maxim of 'seeking out the enemy's fleet' it should be borne in mind: (1) that if you seek it out with a superior force you will probably find it in a place where you cannot destroy it except at heavy cost. (2) That seeing the defensive is the stronger form of war then it is *prima facie* better strategy to make the enemy come to you than to go to him and seek decision on his own ground.[17]

When Corbett wrote this the underwater threat was ill developed and the aircraft merely an unreliable form of observation. The development of these two forms of naval warfare would seem to make Corbett's warning even more true today.

The alternative to battle if one side or the other does not wish to seek it is blockade. Again this is a rather old-fashioned term and NATO prefers 'containment' in its current Concept of Maritime Operations.[18] If the enemy's main forces can be confined to their ports or to restricted areas far from where one needs to sail one's own shipping, then one possesses many of the fruits of victory without the need to take the risks of battle. Blockade, however, consumes much effort over time and its efforts are often far from complete, especially in the era of the submarine and aircraft which not only can keep blockaders from too close a proximity to their own bases but which can evade the blockade more easily than surface ships. Nevertheless, given the geographical disadvantages under which the Soviet Union labours with the various 'choke points' through which its naval forces must pass, various forms of blockade (using mines, sonar arrays, or barrier patrols of various kinds) are useful weapons in the inventory of the West. Submarines operating forward are also potent blockading devices sometimes, as in the Falklands War, having the facility to force an opponent not equipped with ASW to stay in port.[19]

In the days of sail, battle, and blockade were the two principal weapons in the inventory of the dominant sea power. The weaker force had two answers. The obvious one was to attempt a *guerre de course*, literally a war of the chase directed against the other side's merchant shipping. The *guerre de course* is a classic 'sea denial' strategy, i.e. the attempt to deny the enemy use of the sea without necessarily being able to use it oneself. For this reason the *guerre de course* was always regarded as very much a second best by classical maritime strategists such as Mahan and Corbett. It had apparently been ineffective when repeatedly employed by France against Britain. As Paul Kennedy has pointed out, this might not be a totally accurate perception of historical reality. The French *guerre de course* was a very powerful form of pressure on the England of the 1690s.[20] Whatever the power of the *guerre de course* in the age of sail, however, the submarine transformed this strategy into a potential war winner against powers dependent on shipping for vital supplies. Germany came disturbingly close to victory against Britain in 1917 and again in 1942, and the United States laid waste the shipping of the Great East Asia C-Prosperity Sphere in 1944–5.

The only way the *guerre de course* has been successfully countered in the past is by the application of the convoy system, i.e.

The state of the art

the direct escort of groups of ships by warships. Before going any further it must be emphasized that close escort was, and is, only part of the overall convoy 'system'. In addition to the escort, 'warships sent to the convoy assembly ports to escort ships assembled for passage in company to their various destinations',[21] there were also support forces of various kinds designed to give additional protection during especially dangerous parts of the voyage. The forward deployed blockading forces also played a vital role both to contain and survey enemy movements. Nevertheless, if warships were not grouped directly with the ships themselves the system did not work.[22]

The convoy system was almost taken for granted in the great maritime wars of the period 1660 to 1815. Mahan was so fixated with his main fleet actions that he almost overlooked the importance of convoy, but in his study of the war of 1812 he made clear his recognition of its central importance.[23] D.W. Waters, a most articulate and important advocate of convoys, has summed up succinctly the change that occurred in the nineteenth century when the old mercantilist system of state control was replaced by the new doctrines of *laissez faire*. The revolution in economic thought, he argues,

> imposed a revolution in naval thought. The object was no longer to protect the ships plying the routes, but to protect the routes plied by the ships. The key stone of the system of trade defence introduced to effect this was blockade, a measure which had never succeeded in the past and which modern developments in ships and weapons offered small prospect of success in the future. The blockade was to contain the enemy fleet until it offered battle then, as a result of its decisive defeat the world's sea routes were confidently expected to be safe to shipping.[24]

It was also expected that the next war would be short.

There are certain echoes of this thinking in modern naval strategic thought. Those who argued before that things had fundamentally changed were, however, proved wrong. The next war was not short and the blockade proved sadly wanting. Only by introducing convoy was the loss rate drastically reduced to acceptable proportions.[25]

One reason convoy was ignored was its 'defensive' association. Notionally 'offensive' operations seemed much more attractive to self-respecting naval officers. In this context the following excerpt from the 1925 edition of the British Naval War Manual is of interest: 'It is desirable by adopting an offensive strategy to force the enemy to assume the defensive and thus prevent him from attacking our trade. . . . In other words if the enemy attacks trade, an attack

Maritime strategy

upon some interest which he will feel bound to defend may result in the withdrawal of forces from trade attack.'[26] Sadly, in World War II such operations were both impractical and ineffective. Fond hopes that the Germans would not be so stupid as to reintroduce unrestricted submarine warfare (which had brought in the USA against them and ensured their defeat) proved groundless. A partial convoy system was introduced in 1939 but, given the diversion of assets to other priorities, it was far from complete. This allowed the Germans, with their limited numbers of submarines, to inflict almost all their losses on Allied merchant shipping. 'Up to the fall of France U-boats had sunk some 200 British, Allied and neutral merchant ships sailing independently. Of the many thousands of ships sailed in French and British convoys they had sunk fifteen, and had lost eight of the attacking U-boats at the hands of the escorts – an intolerable exchange rate for the U-boats.'[27]

Throughout the war the story was the same, high loss rates among independent shipping, low loss rates and good exchange ratios in U-boats sunk for the convoys. U-boat aces made their names against unescorted shipping, not convoys. At the beginning of 1942, the maturing of the Atlantic convoy system having made life in the open oceans unprofitable, the German submarines were unleashed against the coastal shipping off the USA's eastern seaboard which the Americans refused to convoy until catastrophic losses had been suffered. Some 505 ships were lost west of 40°W between mid-January and the early part of July 1942.[28] Ships safely protected in the open ocean were thrown away in coastal waters. It is interesting, and not a little ironic, that the standard map used by the US Navy today to justify its new doctrine of 'forward operations' quotes the losses of this period.[29] The reason for these losses, however, was not over-commitment to the defence of shipping in home waters, rather it was the virtual complete lack of these defences in a coherent form. Many US and British naval assets were indeed forward deployed, but their contribution to maintaining the flow of shipping was indirect and insufficient.

Again convoy came to the rescue. In July 1942 only 0.3 per cent of ships in US coastal convoys were sunk.[30] The U-boats came back to mid-Atlantic to throw themselves in packs against the convoys. Here the escorts were at their weakest due to the difficulty of providing air cover. There were apparently higher priorities for the necessary escort carriers and the long-range maritime patrol aircraft required. When air escort was finally provided however, and utilization of escorts made more efficient by increasing the size of convoys in accordance with the findings of operational research, convoy escorts and support groups were able to inflict such exchange ratios on the U-boats that Donitz had to call them off: 'their sinkings were negligible, their losses crippling'. German attempts to break back into the Atlantic against a mature convoy system were 'a complete failure'.[31] In the Pacific the success of the American *guerre de course* against the Japanese was in large part due to the latter's failure to come up with a proper convoy strategy.

That convoy should be so effective in maintaining shipping should not be surprising as it combines the advantages of the offensive and defensive forms of war as analysed by Clausewitz and Corbett. Forcing the enemy to come to you means that he must often operate at the end of long lines of communication, in less strength than he might otherwise be able to attain, and perhaps in waters where you can maximize your own strength in relation to his. The defender can thus concentrate his assets where they can be doubly effective – both offensively to mount a crushing counter attack, and defensively to protect from loss the ships whose movement and survival are the object of the operation. The enemy may be deterred from attacking at all and, even if a covert enemy strikes, the 'flaming datum' thus obtained gives away his presence opening him up to counter-attack and destruction if friendly forces are there to do the counter-attacking and destroying.

This brings us to the important paradox that convoy seems often to require large numbers of defensive forces. The argument that these numbers are impossibly high has a pedigree almost as long as convoy itself. Astronomical figures of the total of ships requiring protection were used to delay the introduction of the convoy system in 1917. In fact, given its combination of the best aspects of offence and defence it has historically been the best method of optimizing the effect of relatively few naval assets.

In the Mediterranean in 1917–18 the local commander was unwilling to introduce convoy: he preferred 'an increased and unceasing offensive which should in time enable us to dispense with the convoys and these methods of defence'.[32] Thus the Allied naval forces in the Mediterranean concentrated on the 'Otranto barrage', a combined containment operation and attack at source composed of offensive patrols by surface warships, aircraft, and submarines, bombing of U-boat bases, and the laying of minefields. It sank a single U-boat and conspicuously failed to stop the German and Austrian submarines getting into the Mediterranean. The limited escort allowed to each convoy, usually a sloop and two trawlers for thirty ships, sank eight out of twelve lost in the Mediterranean in 1917 and 1918, and kept the loss rate of Allied ships in convoy at one per cent.[33] To take another example, this time from World

Maritime strategy

War II, between September/October 1943 and the end of May 1944, an aircraft escorting a convoy was ten times more likely to sink a U-boat than one on offensive patrol over the Bay of Biscay.[34] Finally, and most tellingly, in the period after the great withdrawal of the U-boats from mid-Atlantic on 23 May 1943, only a dozen ships were lost from the many convoys that ran through the mid-Atlantic area and none from convoys with an air escort.[35]

It is, of course, true that the technological conditions upon which these performances were based have been fundamentally changed; we shall return to this question in Chapter 10. Nevertheless, the overwhelming weight of historical evidence seems to demonstrate that unless some direct protection is given to shipping then the exercise of command of the sea may be impossible. In other words, the battle around the ships themselves may be as vital to gaining command of the sea as operations elsewhere, perhaps more so. In this sense it is perhaps misleading to refer, as do the classical strategists, to the direct defence of trade as merely the *exercise* of command of the sea. Command is gained elsewhere only to the extent that the battle/blockading forces neutralize the enemy's most capable assets which might otherwise be able to deal with the relatively light and dispersed convoy escorts. The escorts, however, are just as important as the heavier and more spectacular forces. Indeed, recent evidence seems to show that they do a great deal more fighting and their battles are the most important battles *that actually take place*. In the world wars it was around the convoys that Germany's most dangerous naval assets, i.e. those the country was able to use, were sunk. It was, therefore, around the convoys that the decisive Allied naval victories against Germany were won. Defence of shipping via convoy ought, therefore, to be added to battle and blockade as a crucial *offensive* means of gaining command of the sea. Far from being a defensive measure, convoy allows the naval *offensive* to be most efficient and effective, a classic Clausewitzian advantage of the defensive *form of war*. This deserves clearer recognition as a principle of maritime strategy, as it explains why a supposedly 'defensive' strategy turns out to be asset intensive. If one does bring the enemy to battle around a convoy one needs *superiority*, like any attacker, to succeed.

Commander Waters has gone so far as to produce twenty-one 'operational laws' of naval warfare based on historical experience. They may be scenario dependent in respect of the technology available but, using Corbett's rule, they demonstrate 'the normal' from which technological developments may draw us. They are thus worthy of note.[36]

The state of the art

1. The risk to an independent ship of being detected and attacked (R_s) varies directly as the number of ships (N_i) at sea. $\qquad R_s \propto N_i$
2. And directly as the number of enemy submarines (U) at sea. $\qquad R_s \propto U$
3. Therefore combining laws 1 and 2. $\qquad R_s \propto N_i U$
4. The risk to a submarine (R_u) when attacking an independent ship of being sunk by an escort force (E) is zero because $E = 0$. $\qquad R_u = 0$
5. It does not vary with the number of ships attacked (N_i) because $E = 0$. $\qquad R_u = N_i \times 0$
6. The risk of a ship in convoy (R_c) varies inversely as the size (S = number of ships in the convoy). $\qquad R_c \propto \dfrac{1}{S}$
7. And directly as the number of convoys (N_c) at sea. $\qquad R_c \propto N_c$
8. And directly with the number of submarines (U) at sea. $\qquad R_c \propto U$
9. Therefore combining laws 6, 7 and 8. $\qquad R_c \propto \dfrac{N_c U}{S}$
10. Therefore *the larger the convoy and the fewer the convoys, the smaller is the risk to a ship in convoy of being attacked.*
11. In the absence of air escort the loss of ships (L_c) in convoy varies directly as the number of submarines attacking. $\qquad L_c \propto U_c$
12. And inversely with the number of escorts with the convoy. $\qquad L_c \propto \dfrac{1}{E}$
13. Therefore combining laws 11 and 12. $\qquad L_c \propto \dfrac{U_c}{E}$
14. The number of submarines sunk (K_c) by surface escorts when attacking a convoy varies directly as the number of attacking submarines (U_c). $\qquad K_c \propto U_c$
15. And directly as the number of escorts (E) with the convoy. $\qquad K_c \propto E$
16. Therefore combining laws 14 and 15. $\qquad K_c \propto U_c E$
17. For an enemy attacking shipping the yardstick measuring victory or defeat, i.e. the *exchange rate*, is the number of submarines sunk while attacking convoys (K_c) per ship sunk in convoy (L_c). $\qquad \dfrac{K_c}{L_c}$

18. The exchange rate also measures the value of escorts (E) with a convoy. If $L_c \propto U_c/E$ and $K_c \propto U_c E$ then

$$\frac{K_c}{L_c} \propto U_c E \times \frac{E}{U_c} \text{ therefore:}^{37} \qquad \frac{K_c}{L_c} \propto E^2$$

19. The operational value of antisubmarine forces, measured in terms of the exchange rate, is proportional to the square of the number of escorts only when the latter are used as convoy escorts, *not* when they are used in other ways, e.g. on 'offensive' hunting patrol. The E^2 law also applies to aircraft if used as convoy escorts. The value of 'hunting patrols' is measured by a different exchange rate, *independent* ships sunk by submarine (L_i) per submarines sunk by hunting patrols (K_h). 'In each war this exchange rate was ten times as costly to the Allies as the convoy exchange rate (Law 17) . . . The "convoy" and "independent hunting" exchange rates when compared give of course a measure of the efficiency of convoy as opposed to hunting as an operation of war. In each war it was about ten times as efficient as hunting.'[38]

    $$\frac{L_i}{K_h}$$

20. The number of escorts (E) required to protect a convoy varies as rather less than the square root of the number of ships in the convoy (S): Therefore the larger the convoy the fewer the escorts required.

    $$E \propto \sqrt{S}$$

21. Each doubling of inter-ship spacing in convoy reduces the risk of loss by 75 per cent but increases the defence perimeter of the convoy by only about 50 per cent.

One useful way in which the *guerre de course* can be supported by a weaker naval power is to maintain a main battlefleet 'in being' to draw off assets that might otherwise be used in convoy protection. The Germans succeeded in doing this in both world wars. The shortage of destroyers caused by the requirement to escort the Grand Fleet played a crucial role in delaying the introduction of convoy in 1917. Only when it seemed that the United States could provide the requisite escort vessels did Admiralty attitudes to convoy change.[39]

The state of the art

Pioneered by the English Admiral of the Fleet the Earl of Torrington as long ago as 1690 as a way of preventing a French invasion, the 'Fleet in Being' strategy has demonstrated considerable potential at various times. Reduced to its essentials it can be defined as the use of options provided by the continued existence of one's own fleet to constrain the enemy's options in the use of his. It can be used both as a sea denial strategy in preventing the superior enemy transporting his valuable goods or troops across an uncommanded sea, or to help to gain command of the sea for one's self by holding down the enemy's assets to protect what is valuable to him while one's own shipping proceeds with little or no molestation. It is sometimes a difficult strategy to follow as avoiding action is frustrating to one's self, one's crews and one's political leaders. The German Fleet mutinied in 1918: Torrington was dismissed after being forced into premature battle by his Government and had to stand trial for his actions. Happily, both for maritime strategy and the hapless Admiral, he was acquitted, although he never served at sea again.[40]

It is the *passage of shipping* that is the main fruit of command of the sea. This shipping can be commercial or military, and it can cover everything from aircraft carriers to cargo ships. The aim of the passage can be to carry mundane cargoes like coal or oil, or perhaps troops for landing. Alternatively or additionally, the aim might be the application of firepower against some target ashore. Such terms as 'sea communications' or 'sea lines of communication' are dangerous in that they tend to mask the fundamental truth that maritime strategy is about *ships*, defined as the *means of movement* on or below the surface. One does not defend the sea as it is usually a matter of indifference whether explosions take place in empty water. It is hardly surprising in this sense that some land-based Eastern European analysts see NATO's dependence on the sea for supplies as a positive asset.[41] The sea cannot be blown up; but ships can be sunk or disabled.

Western maritime strategy has tended to emphasize the passage of shipping carrying goods and supplies, as it has largely been based on the experience of Britain, the last and greatest of the maritime empires of the 'Age of Vasco da Gama' which ran roughly from 1500 to 1900. Continental maritime strategists have tended to emphasize sea denial like the French *'Jeune Ecole'* of the late nineteenth century. The Germans dallied with Mahan in the early twentieth century but adopted *'Jeune Ecole'*-like 'tonnage warfare' concepts in the late 1930s.[42] The Soviet Union has taken a slightly different approach by emphasizing the utility of 'operations against the shore', both in terms of defending against them and carrying them out. Admiral Gorshkov understandably made much of these in his writings, having organized much amphibious warfare in the southern sector of the Eastern Front in World War II.[43]

Soviet amphibious operations in World War II, although numerous, were relatively short ranged and coastal or riverine in nature. There was nothing to compare with the massive cross-Channel invasion of 1944 or the trans-oceanic amphibious operations in the Pacific carried out by both the Japanese and the Americans. The techniques of amphibious warfare were developed enormously by the United States, in particular with the birth of a whole new family of amphibious warfare vessels. These have made the application of amphibious military power more practical than it has ever been before, although on the other side of the equation the increased scale of coast defences has kept up. Judging the balance of advantage can, therefore, be difficult. The tendency to write off amphibious landings as impractical has been persistent throughout this century but the lessons of World War II and since are that, if sufficient surprise is obtained and/or air power can be brought to bear both to neutralize the inherent advantages of land communications and help together with naval gunfire support to overwhelm the coast defence, amphibious warfare even against significant opposition can still be a practical operation of war.[44] We will return to this matter in Chapter 4.

One of Corbett's most controversial arguments was that sea power allowed the use of limited, but significant, pressure against major opponents in order to force political concessions and achieve the objectives of the war. Corbett quoted the recent Russo-Japanese War as his main example but, as Dr Andrew Lambert has pointed out, a closer examination of the Russian War of 1854–6 would have yielded a more interesting example still.[45] Dr Lambert argues convincingly that the Anglo-French alliance was remarkably successful in exerting maritime pressure against the Russian Empire. This was less in the Crimea where 'a combination of inter allied friction, hesitant leadership and political weakness allowed the operation to degenerate into a long campaign in which two entrenched armies fought out a battle of attrition for a ruined city'.[46] Given the absence of railways on the Russian side, the ability of the Allies to capture Sebastopol, and later Kinburn at the mouth of the Dnieper, was not too surprising. But neither success in the south directly caused Russia to accept the Allies' terms. However, Russia did submit. Dr Lambert may well have identified the reason in British activities, and threatened activities, in the Baltic.

Once Sebastopol had been occupied, the British next planned to unleash in 1856 a 'Great Armament' upon Kronstadt, the fortress guarding the approaches to St Petersburg. The 'Great Armament'

The state of the art

was a massive bombardment fleet of about four hundred steam gunboats and mortar vessels which would attack Kronstadt with mortars, guns, and rockets. The attacking force also included eight ironclad floating batteries and no less than twenty-four line of battleships. With Kronstadt demolished and occupied 'something the responsible officers had every confidence in, St. Petersburg would be open to any insult the British cared to administer.'[47] Sweden might also join the Allies to regain Finland. In these circumstances the Russians gave in and began peace talks during which the assembly of the 'Great Armament' was used by the British 'to back up their tough stand . . ., both to beat down Russian resistance and to impress the neutrals'.[48] Russia was not in danger of complete defeat but the balance of advantage was clearly to submit to British pressure. Dr Lambert concludes that 'The Russian War demonstrated that Russia was wide open to maritime threat.'[49] Both Admiral Gorshkov and his successors, and the authors of the new American 'Maritime Strategy', would hasten to agree that she still is.

Before concluding this survey of the state of the art it is useful to look at what Corbett called the 'theory of the means – the constitution of fleets'.[50] Corbett wisely recognized that the differentiation of the ships of his time into three groups, 'battleships, cruisers and flotilla', while it had 'so long dominated naval thought that we have come to regard it as normal, and even essential', was by 'no means constant.'[51] He perceptively described the evolution of the sailing fleets of the eighteenth century into 'battleships', designed primarily to secure command of the sea, and 'cruisers', faster but less powerful ships, designed primarily to exercise that command by giving direct protection to shipping, and secondarily to support the battlefleet by scouting. The flotilla was the mass of smaller vessels 'for coastwise and inshore work, despatch service and kindred duties'.[52]

Corbett recognized how during his day the classic threefold distinction was being eroded, both by the construction of large cruisers of similar fighting power to battleships, the advent of light cruisers designed primarily for fleet work and the development of the flotilla into the ocean-going torpedo boat destroyer, a vital component of the battlefleet in both offence and defence. Corbett sensed that these developments reflected the dominance of the decisive battle or 'overthrow' school of naval strategy.[53]

As usual, he was right. When the teaching of the latter school proved insufficient to give the Allies complete command of the sea in World War I a whole new class of non-battleworthy 'cruisers' had to be developed, the escort sloops, to accompany the ocean-going destroyers that were also pressed into service to escort convoys against submarines. To a considerable extent the 'mating' of

Maritime strategy

these two types of vessel – the relatively slow 'sloop' optimized for the escort of merchantmen against submarines and later, aircraft, and the fast fleet destroyer, which in order to carry out its fleet defence function in World War II became primarily an anti-air warfare and anti-submarine vessel – has provided the parentage of the modern major surface combatant in its various guises and forms. The ships are, however, primarily classical 'cruisers' in that they are not the primary striking forces of the major navies: that role has fallen to submarine and aircraft. Surface ships are still necessary to screen some of these striking forces, those based in mobile surface ship platforms, i.e. aircraft carriers. Moreover, if it is accepted, as argued above, that the battle around the convoys has become a primary means of gaining fleet command in itself, the assets engaged may well be 'battleships' of a kind. Nevertheless, in that their main role is the escort of higher-value shipping, modern surface warships, 'destroyers' and 'frigates', are classical cruisers first and foremost. Indeed, the larger examples have taken over the name as it was vacated by the last of the large, gun armed 'neo-battlecruisers' of the first half of this century, ships designed for surface warfare and combat with their own kind. The large maritime patrol aircraft, designed primarily for anti-submarine work, might also qualify as latter day 'cruisers' of the air.

What has become of the battlefleet? As stated above, it has taken to the air and gone beneath the waves. It was soon clear that aircraft had the *potential* to sink anything that floated, but the air enthusiasts overstated, sometimes grossly, the capabilities of the aircraft available during the two decades after World War I. The Naval Staffs of the time were right to stick to the large gun-armed battleship as the core of their capability to sink the heaviest and toughest ships available to the other side. However, the richest and least strategically over-committed of them, the Americans and Japanese, were also able to develop aircraft, both carrier and sometimes shore based, as complementary striking arms. About 1940–1 the aircraft did overtake the battleship with the advent of the high-performance torpedo bomber, supplemented by high-performance dive bombers with heavy bomb loads. In the Pacific, aircraft soon asserted their position as *the* dominant strike arm against even the largest enemy surface warship. The largest battleships ever built, the 70,000 ton (full load) behemoths *Yamato* and *Musashi*, were overwhelmed by American carrier-based aircraft despite their massive AA gun armaments and their being able to manoeuvre in the open sea. The former blew up after being hit by perhaps thirteen torpedoes and eight bombs, the latter by twenty torpedoes and seventeen bombs.[54] Although land-based aircraft

The state of the art

could be as effective, as when the Japanese 11th Air Fleet disposed efficiently of the battleship *Prince of Wales* and battlecruiser *Repulse* in December 1941, carrier-based aircraft tended to be more useful and flexible under maritime conditions. Hence the large strike carrier become the new 'battleship', one capable of fighting a three-dimensional anti-surface, air and underwater battle. Sadly, for the rest of the world's navies only the United States is now capable of affording such mighty ships. The rest must make do with smaller carriers (better described as large 'cruisers') or land-based maritime strike aircraft. The Soviet Union was forced to use missile technology to substitute for non-availability of proper carrier aircraft. From the late 1950s she produced what were in effect 'battlecruisers' to complement her primary submarine and land-based air striking forces, culminating in the monsters of the 1970s and 1980s, some relying on a mix of aircraft and missiles and others on missiles alone.

In one important way the modern attack carrier is also better described as a 'battlecruiser' than a 'battleship' in that she relies overwhelmingly on striking power to protect herself, not armour protection. In this she is very like Admiral Fisher's original conception of a battlecruiser, a ship that could strike at enemy vessels from ranges where they could not hit her. Sadly, Fisher's assumption of British gunnery superiority was let down by his subordinates and successors refusing till too late the fire control system that might have given their ships this ability. Nevertheless, in a sense the erratic British naval genius of the twentieth century was vindicated by the replacement of the traditional battleship by the carrier as the dominant surface warship.[55]

In 1919, shortly before the end of his tempestuous life the following year, Fisher summed up naval development as he saw it: 'The greatest possible speed with the biggest practical gun was up to the time of aircraft the acme of sea fighting. Now there is only one word – submersible.'[56] As usual Fisher overstated his point – he may well indeed have been considering semi-submersible gun armed battlecruisers rather than submarines proper – but there was a grain of truth in it nonetheless. By the time he wrote the above the submarine had passed from being a mere coast defence flotilla vessel to a powerful commerce-raiding cruiser. Attempts by the British, and later the Japanese, to make it into a fleet cruiser, however, were less than successful, and only the Americans proved able in World War II to use submarines as effective striking forces in fleet actions as a kind of 'advanced cruiser line', notably at the great victories at Philippine Sea and Leyte Gulf in 1944. Since 1945, however, the submarine has grown into a full-scale strike arm symbolized by the

Maritime strategy

'Dreadnought' names given to Britain's early nuclear-powered submarines. Interestingly, the Americans, who in the 1920s and 1930s built heavy cruisers, unnecessarily large in British eyes, primarily as fleet 'battlecruisers', have chosen cruiser names for their submarine strike force.

The fact that the primary maritime striking force of the world's major navies is split between air and submarine platforms makes one over-used term obsolete – 'capital ships'. Oddly enough this term came into vogue in the early years of the twentieth century as an umbrella term for large gun-armed battleships and battlecruisers just as these vessels were increasing their dependence on lighter screening forces. Nevertheless, it did have some meaning as 'the unit of absolute power . . . unsinkable except in combat with a superior force of similar ships'.[57] No vessel is in such a happy position today. Submarines can torpedo carriers, carriers can take on submarines. Equally submarines cannot shoot down aircraft. Even old World War II destroyers dropping depth charges can neutralize the activities of grandly named nuclear submarines. There are no 'capital ships' now and modern naval warfare is a complex three-dimensional kaleidoscope. However, before going further into exploring its nature we ought to discover what is being fought about: what uses of the sea are worth all the trouble? And how important is the sea in the violent resolution of conflicts caused ashore?

Part II
The uses of the sea

Chapter 3

Economic uses of the sea

The main traditional use of the sea is as a medium for transport. Despite the advent of the transcontinental railway and the transoceanic aeroplane the sea still dominates international transport movements. Seaborne trade, as we have already pointed out, accounts for over 80 per cent of international trade by volume.[1] If weight is the criterion then the importance of water transport increases still further. 'No less than 95 per cent of all trade that crosses frontiers is waterborne, including inland. Over 99.5 per cent by weight of all trans-ocean cargo is carried in ships, under half a per cent in aircraft, and there is no viable alternative mode for most of it.'[2] World seaborne trade is rising again, 3,362 million tonnes being carried in 1986, some eight times the amount carried in 1946. The trend is upwards despite only 3,090 million tonnes being transported in the trough of the recession in 1983. The previous peak was 3,606 million tonnes in 1980.[3]

Crude oil and petroleum products provide by far the largest single category of seaborne trade, an estimated 940 million tonnes of the former and 310 million tonnes of the latter being carried in ships in 1986.[4] The Middle East dominates the export of oil by sea with about half the world's exported oil originating there. Western Europe and Japan are the main customers, over a third of the former's total oil imports and no less than two-thirds of the latter's being Middle Eastern.[5] All of this must move in tankers, the former (mostly) round the Cape and through the Suez Canal, the latter across the Indian Ocean and then through the many Straits leading into the South China Sea.[6] Large quantities of oil are, however, carried by sea within Western Europe, from North Africa to Western Europe, West Africa to Europe and North America, Soviet East Asia to Japan, Alaska to the continental United States (some through the Panama Canal), and from the Caribbean to North America and Europe. The complete pattern of the world seaborne oil trade is shown on Figure 3.1.

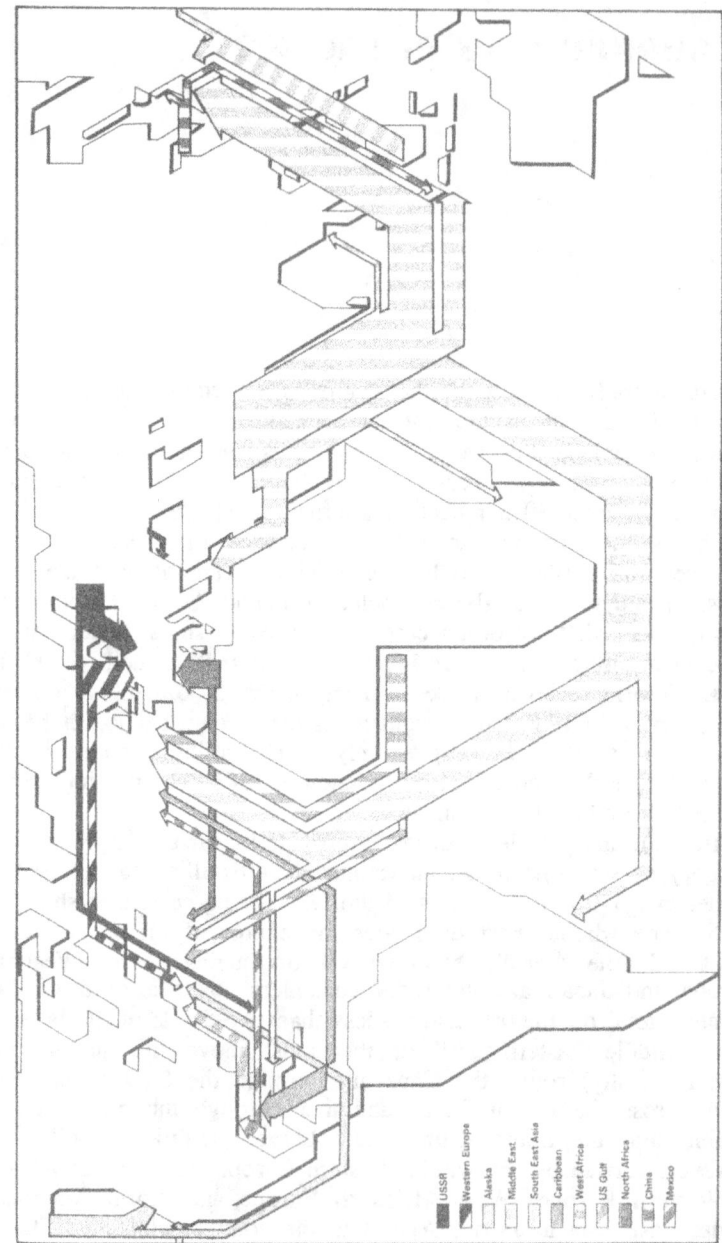

Figure 3.1 Main oil movements by sea, 1987

Source: *BP Statistical Review of World Energy*, June 1988, p. 19.

As noted above, most oil is carried as crude. This allows refineries closer to the markets to obtain the best blend of crudes for their production. The trend towards long hauls of crude caused tankers to increase steadily in size from 1950 onwards. By 1959 the first 100,000 deadweight ton ship had been built and over the next decade the very large crude carrier (VLCC) 200 to 300 thousand deadweight tons (dwt) was born. With the closure of the Suez Canal after the 1967 Arab-Israeli War forcing long voyages from the Gulf round the Cape to both Europe and the United States, and the very high profits obtainable as oil demand steadily increased, large numbers of VLCCs were constructed. By 1973 the ultra large crude carrier had arrived, the monster of half a million deadweight tons. At this point economic conditions sharply changed. Oil prices increased, demand diminished and, not least, the large oil companies lost their dominance in the oil market place. The latter factor 'reduced the average size of their oil parcels shipped by sea, increased the number of tanker routes and increased the number of tanker charterers'.[7] Demand for VLCCs slumped by half between 1978 and 1982, and few were constructed. The reopening of the Suez Canal and its deepening to allow the smaller VLCCs through also has put a new premium on moderate dimensions for carrying crude oil. Petroleum products like petrol, diesel oil and paraffin have always been carried in much smaller ships, nowadays of around 25,000 dwt and smaller ships still are used for coastal and riverine traffic.[8]

Another major bulk hydrocarbon traffic is liquified gas. Taking liquid petroleum gas (LPG – butane and propane) from the oilfields, where it was often flared off as waste, to consumers in the developed world began in small ships in the 1950s. For larger ocean-going LPG carriers refrigeration began to be used to help store the volatile cargoes and this led the way to ocean transportation of natural gas (LNG – mainly methane). Japan, especially, requires large amounts of LNG to come by sea and this demand is fulfilled by impressive and expensive specialized LNG carriers capable of carrying 125,000 or more cubic metres of volatile cargo at temperatures below its boiling point of $-161\,°C$.

Moving now to solid bulk cargoes, the most important is iron ore, of which 304 million tonnes, 9 per cent of total goods, were carried by sea. About a third of world production is carried by sea, and the ore producing areas of Sweden, Australia, Canada, Brazil, Venezuela, India, and West Africa are linked to the production centres of Japan, West Germany, the United States, the United Kingdom, Belgium, France, and Italy. All the latter producers use large amounts of imported ore. Although iron ore can be carried in non-specialized tramps, most is transported in bulk carriers of around 35,000 to 45,000 dwt tons. For long voyages, notably those to Japan, the bulk carriers used may be quite large – in the 100,000 to 150,000 dwt bracket. Some ships in this category are so-called 'combination carriers', doubling as transporters for oil and ore. Such ships, however, tend to be expensive and awkward to operate, factors militating against their flexibility.[9]

Almost as much coal is carried by sea as iron, some 268 million tonnes in 1986, 8 per cent of total sea trade. This is required by steel industries with insufficient supplies locally, and also for electrical power generation. Japan and Europe are the major importing areas: the former consuming a third of world imports. Major producers for seaborne distribution are South Africa, North America, India and Australia. Eastern Europe also exports significant quantities of coal by sea through the Baltic. Coal is carried in bulk carriers in the 35,000 dwt bracket. Ships of up to twice this size can be used – the largest that can transit the Panama Canal – a significant consideration with traffic to be carried from the Eastern Seaboard of North America to Japan.[10]

Another important dry bulk cargo is grain, 160 million tonnes of which were shipped by sea in 1986, 5 per cent of total trade by weight. Depending on demand, North America and Australia export all over the world. Standard bulk carriers are the normal grain carriers, although even tankers are pressed into service at times. Grain is also a staple cargo for the small tramping vessel. The above cargoes total 270 million tonnes, and amount to 8 per cent of total trade. Moving on to other bulks: bauxite travels from Australia, West Africa, and the Caribbean to Europe, Japan, and North America; phosphates from the United States, the Pacific islands, and Morocco to Europe, Japan, and Australasia; manganese ore from Brazil, Africa, India, and Australia to Western Europe, North America, and Japan; copper from Canada, the Philippines, and Papua New Guinea to Japan; nickel from New Caledonia, Indonesia, and the Philippines to Japan; zinc from Canada, Peru, and Australia to Europe and Japan; chrome from South Africa to Japan and elsewhere; sulphur from North America and Poland to all over the world; and sugar from South America, the Caribbean, Africa, Australia, the Pacific islands, and various parts of Asia to North America, Europe, and Japan (and from France to Africa and from Asia and Cuba to the USSR). The medium-sized bulker is the rule for most cargoes, although large combination carriers and tramps are used for some, e.g. sulphur and phosphorus. More complex chemicals are carried in small (25,000 ton) specialized tankers. These ocean-going tankers cover two main routes, from the Gulf of Mexico to Europe and from

Economic uses of the sea

Newfoundland to both Europe and Japan. As with petroleum products, smaller chemical tankers still are used in coastal waters.[11]

Exactly a third of all seaborne trade by weight in 1986, 1,110 million tonnes, was classified as 'other dry cargoes', e.g. fruit, meat and manufactured goods. Once the preserve of the developed world, increasing standards of living in the developing countries are helping maintain demand for such items. Dry cargo ships can be liners, operating regular routes to a fixed schedule, or tramps sailing where the cargoes can be found. The traditional type of vessel for both types of service used to be the 'break-bulk ship' with holds loaded with two to five ton lots of cargo by slow, labour intensive means through hatches into the hold. As, after 1945, labour increased in cost and ships, cargo liners especially, got faster, so the relative cost of the masses of dockers required and time spent idly in port increased. Ships were often tied up for more than half the year. Very large break-bulk ships were not possible because of the long turn round times required.[12]

The answer was 'unitization', 'the breaking down or, alternatively the syntheses up, of goods into standardised units that can be handled by mechanised and possibly automated equipment'.[13] The most important form of unitization is the use of the standard ISO container, 8 × 8 feet in cross section and 10, 20 or 40 feet long. The 20 foot container, the usual standard, has a maximum payload of about 18.5 tonnes.[14] The advantages over break bulk are obvious: 'As break bulk, drums of paint, wood bales, cars, crates of canned goods, cement bags, machine tools and so on have to be packaged, crated and handled in different ways, and stowed differently in the ships themselves, but all can go in containers which can be handled in one standardised manner no matter what is inside them.'[15] Containers require specialized capital intensive port facilities, capable of handling ten times the cargo with only a third the labour.[16] Between 1965, when the ISO container was born, and 1980, dry cargo sea transport in the developed world was transformed in 'even more of a revolution than that in break-bulk cargo following the introduction of really efficient steam engines'.[17] In addition to the fundamental advantages of lower port costs and greater time at sea, pilferage was significantly reduced (10 per cent) and crews were greatly reduced in size (90 per cent).[18]

Container ships have rapidly evolved through several generations. They were always relatively large given the increased volume required. A typical ship of the 1970s would be of about 20,000 dwt tons with a capacity of 1,500 twenty-foot equivalent units (teu) and a speed of 21.5 knots. During the 1970s capacities increased to 2,000 – 2,500 TEU and speeds to 26–30 knots. The rise in the oil price soon

The uses of the sea

undermined the economies of the fast container ship, and at the beginning of the 1980s re-engining with slower and more economical engines, usually diesels, became the rule. Economical cruising speeds are now 18 knots. Container ships of the 1980s vary from small ships of under 15,000 dwt tons to the huge 4,380 teu, 58,000 dwt giants used by United States Lines on their round the world route. The limiting factor on such vessels is the requirement to transit the Panama Canal.[19]

Even container ships suffer from supply outpacing demand and a growing tendency is to build ships in the 15,000–45,000 dwt bracket that can carry either containers or bulk cargoes. Such ships can move between trades when squeezed by problems of over-capacity in one or the other. One example of such a ship is the British designed 23,000 dwt Ecoflex type designed to carry general cargo, grain, ore, or containers. The Scandinavians have developed a type of ship known as the 'Bo-Ro' which combines this flexibility with roll-on roll-off facilities. A Bo-Ro's voyage pattern demonstrates its enormous versatility:

> Loading 5,000 tonnes of cargo in the Far East, including mobile cranes, motor cars, containers, etc., at rates varying from 300 tonnes per hour for containers to 1,000 tonnes for the vehicles, the ship was unloaded in North West Europe. There nearly 10,000 tonnes of oil cargo were lifted for a short coastal feeder trip. After discharging this, 10,000 tonnes of forest products were loaded on trailers at nearly 700 tonnes per hour. This was in turn unloaded in Belgium where 700 motor cars were taken on, together with 9,000 tonnes of fuel oil in Amsterdam and the whole of the cars and fuel cargo delivered to Scandinavia.[20]

Roll-on roll-off (Ro-Ro) facilities are space intensive but repay the investment in terms of ease of loading and unloading. The technology was pioneered by the landing vessels of the World War II. The first Ro-Ro ships were in fact converted British tank-landing ships used on short sea operations in British waters from 1946 onwards. The terminals used were also wartime in genesis – pontoons and bridges used in the Mulberry harbours. Specialized Ro-Ro cargo ferries followed and by the 1970s ocean-going Ro-Ro ships of 35,000 dwt were in service.[21] Some of the most impressive Ro-Ro cargo liners today are the 36,500 dwt combined Ro-Ro and container vessels used by the Atlantic Container Lines consortium on the North Atlantic run. With a capacity of over 2,000 teu and over 600 vehicles they are the Atlantic liners of today.[22]

Modern ships such as these tend to dominate the trades of the developed world but the break-bulk vessel is far from dead. There are still more ships of this type than any other in the World's trading fleet. Where high labour costs can be afforded for various economic reasons, e.g. in the developing world or in the Socialist countries, it remains viable, if not essential, for the type of port facilities available. Container handling facilities are, however, spreading. Ports with new installations in recent years have included Bombay, Calcutta, Cochin, Colombo, Curacao, Karachi, Madras, and Rangoon, as well as Chinese, Egyptian, Libyan, and Thai ports.[23] Although old break-bulk vessels are still so cheap as to be not worth replacing with new construction, one would expect container vessels with cranes and holds to slowly displace them as the years go by and containers become even more widely used. The current design of Soviet general cargo vessel is a 13,500 dwt ship, capable of carrying 320 containers in its hold, 68 more containers on deck, loads unitized in pallets, or even cargoes of bulk grain.[24]

Many old break-bulk vessels are capable of carrying refrigerated cargoes, largely food. The 'reefer' trade has kept relatively healthy, thanks to the rising standard of living in the developed world and the expanding demand for new and different kinds of fruit. The Middle East has also become a major food import market with the rising standards of living there caused by oil price rises in the last decade. Container ships, normal liners with provision for refrigerated cargo, have taken some of the reefer traffic but the seasonal and/or more specialized facilities (e.g. bananas) are becoming the preserve of modern specialist reefers designed to carry their cargoes in pallets rather than containers.[25]

Another kind of specialized 'unit load' vessel, and one of the few types of ship for which demand remains good, is the vehicle carrier. The Asian motor industry provides a demand for significant fleets of such vessels, sometimes owned by the motor manufacturers themselves, each one carrying over 2,000 cars at up to about 18 knots from the factories to Europe, North America, or elsewhere. The vehicles are simply driven on and off the ships.[26]

A final and spectacular form of 'unit cargo' vessel that must be mentioned is the large carrier ('lighter or barge aboard ship' vessel – LASH or BASH). These are ships that carry lighters or barges to be floated or lifted on and off the ship at each end of the voyage. The first appeared in 1969 and was soon followed by ships designed to carry both containers and barges. Such ships have a number of advantages, notably the ability to land cargoes without complex harbour facilities and to send barges directly into inland waterway systems. The early ships were all American, except for one Dutch vessel, and did not prove as popular as expected, possibly due to the availability of Ro-Ro and container vessels at all major ports dealing

The uses of the sea

with likely cargoes. Nevertheless, the concept was taken up once more in the late 1970s in both West Germany, with its extensive inland waterways, and the Soviet Union. The latter has a less developed port infrastructure and a tendency to use merchantmen for military purposes: both considerations made LASH vessels attractive. A number have been built in Western yards for the USSR, both ocean-going ships and smaller feeder vessels for operating at the ends of the routes of the larger ships and for relatively short sea voyages. Soviet yards were used, however, for a new class of impressive 41,000 dwt ships designed for arctic operations. These Aleksey Kosygin class ships can carry 82 barges or almost 1,500 TEUs and are used on the run between Vladivostok and the mouth of the Yenisey river in the Kara Sea. This Arctic route is the Soviet Union's only way of providing heavy transport facilities to its Northern Siberian lands; given the ice conditions, nuclear power would have advantages. The Kosygins are diesel propelled but it now seems that the slightly smaller successors of the Sevmorput class (32,000 dwt and 73 barges/1,300 teu) are nuclear powered. This form of propulsion has not found favour in the West for merchant shipping, despite various experiments. It is only practicable in Soviet-style economic conditions and even then for special purposes like Arctic navigation. The USSR has a powerful fleet of icebreakers to break passages for its more conventional merchantmen: some of these icebreakers are also nuclear propelled, the first, the *Lenin*, was in 1957 the world's first nuclear-powered surface ship. She suffered a serious nuclear accident nine years later, but now appears to be back in service.[27]

The Soviet Arctic sea route is perhaps the most spectacular example of coastal and short sea trades. All over the world considerable volumes of traffic are carried on short voyages by ships, many of which can penetrate far up river or internal canal systems. Modern coasters must be able to carry containers and the 'open hatch' vessel has become popular. Ro-Ro is also a useful asset when time in port is at a premium. Coasters are usually quite small, varying from under 500 to over 2,500 dwt, but about 1,100 dwt is the minimum for profitability, in Western Europe at least where the German Federal Republic is now perhaps the leading coasting nation.[28]

Short distance sea traffic also remains competitive in the transport of passengers, especially if they are accompanied by their cars. Ferry services using Ro-Ro techniques play a crucial role in the transport systems both of Western Europe and Japan. An important route, although one with freight and military cargoes in mind, is the new combined 'rail and car ferry' route between Kleipada in the Soviet Union and Rugen Island in the DDR. This will have a

Economic uses of the sea

capacity of over 5 million tons per year by 1990 and will provide more secure transit, avoiding the Polish railways. Ferries are becoming progressively larger as ferry operators try to meet increased freight demand, maximize general economies of scale, and give greater passenger comfort, especially on overnight sailings, all in a highly competitive environment. In 1987 British Shipbuilders delivered a Ro-Ro passenger ferry of no less than 31,000 gross registered tons, the largest passenger ship built in the United Kingdom since the Queen Elizabeth II in 1967.[29] Their owners hope these new, large ferries will help stave off such threats as the Channel Tunnel. Sea transport will, however, remain the only means of movement in relatively remote areas, e.g. in northern Norway. New types of small craft, e.g. catamarans and hovercraft, found their main commercial niche in fast, short distance coastal services where land alternatives, for various reasons, were unavailable.

The larger passenger liners now remain almost purely as cruise ships, although sometimes a limited regular passenger service is offered when ships sail to and from their cruising grounds. One significant by-product of the end of the ocean passenger liner as a means of transport rather than a means of recreation has been the drop in speed of such vessels. Once some of the fastest ships afloat with speeds of 28 knots or more, liners now rarely exceed a more economical 22 knots. The world's largest liner, the *Norway*, had her speed reduced from 30 knots to 18 knots when she was refitted for cruising, minus two of her original four turbines. Converting a ship for cruising is not just a matter of reducing the speed, the whole layout must be altered to optimize its ability as a floating hotel, swimming pool, and sun lounge.[30]

In 1983 the United Nations listed the percentage deadweight tonnage of the world's mercantile fleet of 37,000 ships as 44.1 per cent tanker, 24.7 per cent bulk carrier, 16.5 per cent general cargo, 7.1 per cent combined bulk carrier/tanker, 2.1 per cent container ship, 0.5 per cent vehicle carriers, 0.4 per cent passenger ferries, 0.1 per cent barge carriers, and 1.2 per cent others. The total tonnage of 686 million dwt was 47 per cent owned by the developed market economy states, 29.1 per cent open registry (i.e. 'flag of convenience') countries, 15.3 per cent developing countries, and 7.9 per cent Socialist nations. The figure of about 37,000 was an increase since 1971 of about 4,000 ships.[31] Since then numbers have fallen again. On slightly different criteria, cargo ships and tankers of over 100 grt, the number of ships in the world trading fleet has declined from 38,307 of 673 million dwt on 31 December 1983 to 36,454 ships of 626 million dwt on 30 December 1986.[32]

The uses of the sea

The breakdown of the world trading fleet is as follows:

	Number	million dwt
Container Ship	1,053	21
Ore/Bulk Carrier	4,945	195
Combination Carrier	363	39
Other Dry Cargo	21,833	109
Tanker	7,495	252
Liquified Gas Carrier	765	10

As for ownership, the top ten merchant fleets by flag on 31 December 1986 were as follows:[33]

Country	million dwt	world percentage
Liberia	97.8	15.6
Panama	69	11
Japan	55.8	8.9
Greece	46	7.3
USA	25.1	4
USSR	23.4	3.7
Cyprus	23.4	3.7
China	17.9	2.9
Hong Kong	13.7	2.2
Philippines	13.7	2.2

Hong Kong is a UK Dependency using the Red Ensign. This flag is worn by ships of a number of different registries. If all Red Ensign ships were counted (notably the United Kingdom itself with 11.1 million dwt tons, Bermuda, Gibraltar and the Cayman Islands) the fleet would move above the USA to number five. Except for the United Kingdom these Red Ensign registries are about as 'open' as the classic 'flag of convenience' states.

'Flags of convenience' proper are those of nations such as Liberia, Panama and Cyprus, who open their registries to owners who wish to employ lower-cost crews than would be possible under national regulations, to avoid paying tax to the national government of the 'beneficial owners' and to avoid some, if not all, regulations for safety, etc. The Red Ensign is the only open registry that can offer naval protection. Flags of convenience are unpopular with the developing countries as they feel that those developing nations which are flag of convenience states do not receive sufficient benefit themselves and that the system provides unfair competition for the fleets owned and flagged by developing countries.[34] The main beneficial owner of flag of convenience shipping has been the United States.

Economic uses of the sea

In 1981, of a total fleet over 100 grt of almost 2,000 vessels, over 550 were Liberian flagged and over 300 Panamanian. More deadweight tonnage, by over a third, flew the single star flag of Liberia than flew the fifty stars of the USA.[35] The league table for current beneficial ownership in dwt is:

1. Japan   89 million tons
2. Greece   71 million tons
3. USA   65 million tons
4. Hong Kong   45 million tons
5. United Kingdom   33 million tons
6. USSR   29 million tons
7. Norway   24 million tons
8. China   18 million tons
9. The Philippines   15 million tons
10. South Korea and Singapore   13 million tons each

Several countries crowd into the 8–12 million bracket: Italy, West Germany, France, Spain, Taiwan, India, and Brazil.[36]

The merchant fleets of the less developed countries are being helped by the 'UN Convention on a Code of Conduct for Liner Conferences', which intended to put trade into ships wearing the flags of the states from whose ports the cargoes originate. This is one part of a general pattern of political intervention in the merchant shipping business, especially the liner trades. According to one British study, up to 20 per cent of the world's liner trade is affected by cargo reservation of one kind or another, about 10 per cent or slightly over of world trade in general. Other ways of helping national shipping are by preferential berth assignment and demanding non-national ships to possess especially elaborate documentation. The United States does not live up to its free market ideology in its merchant shipping policy with cargo-sharing arrangements with Latin American countries and the forcing of all trade between American ports into American ships. The USA also pays considerable operating subsidies to its shipping lines. The Socialist countries, notably the USSR, do not, of course, work to Western notions of profit and loss: this allows rate cutting below the agreed 'western' Liner Conference figures although service can also be correspondingly poor.[37]

The Socialist countries play an important role in the second, even older, form of sea use – fishing. This is probably more important today than ever, although the world fishing industry has changed greatly in structure over the last decade with the shutting out of the Western European fleets from their traditional fishing areas in the

The uses of the sea

Atlantic that now form part of the other nations' Exclusive Economic Zones (EEZs); 95 per cent of the world's catch is taken within 200 miles of the shore.[38] Although these Western nations still own substantial fishing grounds of their own they also import substantial, perhaps greater, amounts of fish from other countries, often straight from foreign fishing boats. The United Kingdom caught 832,500 tonnes of fish in 1985 of which it exported 304,100 tonnes; UK fish imports, however, totalled 891,200 tonnes.[39]

Fish provides 25 per cent of the world's supply of animal protein and is especially important to the less developed countries. In these nations it provides the only supplement to rice or maize. 'If countries are ranked by reliance on animal protein derived from fish, 39 of the first 40 places are occupied by developing countries.'[40] Among the developed states both Japan and the USSR rely heavily on fish for food. Both countries have for various reasons a deficient agricultural base and need protein from the sea. The Japanese diet obtains half its protein from fish and the Soviet Union is trying to stimulate domestic fish consumption to 40lb per head per annum.[41] Fishing is also important, especially in the Soviet Union, as animal feed and fertilizer.

Fishing is carried out in relatively restricted areas of the world's oceans. Almost all the world's catch comes from about a third of the world's sea area. Most fish (about 75 per cent) comes from the Sub Arctic and Continental Shelf areas of the North Atlantic and North Pacific, and the waters to the west of Africa and the Americas, although other significant areas are the South Atlantic, the northern Indian Ocean, the seas around South East Asia, and the area between the Philippines through Indonesia to Northern Australia.[42] Given their state of economic development and their demand for fish, it is not surprising that the greatest fishing nations of the world are Japan and the USSR, which caught 11.4 million and 10.6 million tonnes respectively in 1984. Both use fish factory ships to give their fleets the capacity to range over the world's oceans, the Soviets having to make the greater effort due to their more restricted sea access. Agreements with distant states provide both access and facilities, although in certain areas fish has to be purchased from local catchers. Both the USSR and Japan have explored new types of catch, for example the Arctic krill, used as a form of animal feed, although finding markets for this product has not proved as attractive as originally hoped.[43]

Smaller countries have invested in distant deep-sea fishing industries too. South Korea and Taiwan follow Japan's model. South Korea has a substantial catch of 2.65 million tonnes gained from the South West and Eastern Pacific and Indian Ocean, as well as waters

Economic uses of the sea

closer to home. Communist states with domestic difficulties of agricultural productivity find distant water fishing attractive; Cubans, Poles, Bulgarians, and East Germans are to be found in the South Atlantic and round Cape Horn in the South Eastern Pacific. In the opposite direction, China, already the world's largest producer of fresh water fish (almost a million tonnes) has also begun to spread over the Pacific and round Cape Horn. Her sea catch totalled 6.78 million tonnes in 1985, against 4.77 million tonnes for the United States. Other countries with substantial sea fisheries of over 2 million tonnes per annum in 1985 were: Chile (4.8 million tonnes – and the world's greatest fish exporter, 1.3 million tonnes), Peru (4.2 million), India (2.8 million), Norway (2.1 million), Thailand (2.1 million), and Indonesia (2.07 million). Between one and two million tonnes were the Philippines, Denmark, North Korea, Iceland, Spain, Canada, and Mexico. Brazil and Ecuador were both getting close to the million mark.[44]

Although the post-1945 era saw the growth of 'industrial fishing' using large self-contained fish factory ships, the legal restructuring of the industry to allow local states greater control of their coastal waters has restored the balance in favour of smaller craft. Now, however, the large vessel is acquiring new uses. Previously industrial fishing was largely to obtain fish meal for agriculture. In the 1980s, however, fish processing techniques are making possible human consumption of otherwise unacceptable species. Pioneered by the Japanese, this technique of producing what is effectively synthesized shellfish has led to a resurgence in large trawlers capable of taking in 400 tonnes of various fish at a time and freezing it. Fish processing will, it is hoped, take the pressure off some more traditionally acceptable species and allow the exploitation of abundant species of ocean fish that have not previously been considered good for anything but animal feed and fertilizer. By such means the world's annual fish catch may be more able to keep up with population increases without over stressing the oceans' fish stocks.[45]

One form of 'fishing' which may become extinct because of environmentalist pressures is whaling. Since October 1985 there has been an official ban on commercial whaling, although Japan, the USSR, South Korea, Norway and Iceland all continued to catch whales. Japan stopped commercial whaling in 1987 and Norway announced that she would stop commercial whaling by 1988. But both countries intend to continue with whale catching for 'scientific research purposes', as does South Korea. Sea-dependent Iceland is the major exploiter of this loophole. Having officially stopped 'commercial' whaling in 1985, about 120 whales were caught in 1986 in a 'research programme' that looked very like 'a front for continued commercial exploitation'; 1987 saw economic pressure from the USA that led to a deal allowing a limited catch on condition Iceland clears further whaling with the International Whaling Commission.[46] It may take some time for the final remnants of this traditional form of maritime hunting to disappear, although it is now of very marginal economic importance.[47]

As well as animal products the sea now provides substantial quantities of minerals, notably oil and gas. Offshore hydrocarbons have transformed the economies of even well-developed industrial states. By 1980, when offshore oil production accounted for 20 per cent of total world output, the United Kingdom, totally dependent on foreign oil supplies ten years before, was producing 1.65 million barrels a day, second only to Saudi Arabia in offshore oil production.[48] In 1985 the United Kingdom produced almost 50 per cent more crude oil than she consumed in 1986; petroleum and petroleum products accounted for over 11 per cent of the UK's exports, although her refineries still required significant amounts of overseas oil for an optimum range of products. Indeed, a surprising 5 per cent of total British imports remained petroleum and its products.[49] Nevertheless, North Sea Oil has been a very significant cushion for the decline of the UK's traditional industrial base.

As well as oil, gas for fuel has become a major product of the sea. In Western Europe the United Kingdom, the world's largest offshore gas producer outside the USA, has replaced traditional coal gas completely with the products of the North Sea. Other major gas producers are Norway, Abu Dhabi (UAE), the USSR, and The Netherlands.[50]

The level of demand and price will dictate the extent of oceanic hydrocarbon extraction in the future. The persistence of the boom has been more fragile than once thought, and even in the longer term alternative energy sources (e.g. coal) may well prove cheaper and more convenient. Similar factors will also govern the viability of the maritime extraction of non-fuel minerals. Dredging of useful minerals in coastal areas has a long history, as does mining out from the coast. Oil bearing silts and brines have been found in the Red Sea but their exploitation is expensive. The geographically limited supplies of manganese may, however, put a premium on the collection of manganese nodules from the deep ocean floor. The technology is becoming available but the profitability of the operation remains questionable, and debate over the ownership of seabed resources militates against confident investment. The United States currently argues that the interests of its technologically advanced deep-sea extraction industries are best served by refusing to accede to the current law of the sea treaty. It fears the assignment to an international authority of valuable resources to which it would be convenient for the United States to have access, and which it is perhaps the only country in a position to exploit. An international authority would reflect the interests of existing land producers of minerals in the Third World. One suspects, however, that the legal uncertainties involved in this particular American exercise in unilateralism will hinder the expansion of marine mineral extraction more than it will help it. It may also find that unexpected competition from international consortia, possibly joining developed with less developed states, makes a totally free market less attractive.[51]

It can be seen from the above, therefore, that the sea remains a crucial element in the world economy. It remains its main form of international transport and it is a major source of food. It provides significant quantities of energy and useful reserves of minerals. The extent to which these economic forms of sea use may cause conflict will be addressed in later chapters. Now we must move to examine the ways the sea can be used as a military medium in conflicts caused both at sea or, more usually, on land.

Chapter 4

The military medium

The various economic uses of the sea provide mankind with assets worth defending and fighting about. Nevertheless, the oceans also provide an important medium for the application of military power, if not for its own sake, then to achieve political objectives that might have little connection with the sea itself. The seas provide the most easily crossed two-thirds of the earth's surface and man has always used them as a way of gaining access to his opponent. With the development of new kinds of longer-ranged weapons systems the use of the seas to 'project power' has become even more important than it has ever been. Indeed, as we have seen, this rather than the protection of economic sea use is probably now the most important function of naval forces in the late twentieth century.

As Mahan memorably put it, the sea remains 'a wide common, over which men may pass in all directions'.[1] The growing moves to enclose the sea for economic purposes have so far not limited very significantly the capability of nations to use the high seas to pass their military forces across the globe. They must still use the sea if they wish to apply military power against distant states. Despite the advent of large transport aircraft of the American C-141 and Soviet Candid categories with payloads in the 40 tonne class, and the even larger aircraft in the 80–125 tonne class, the US C-5 and the Soviet Cock and Condor, the capacity of aircraft to move large amounts of heavy equipment remains quite limited. One of the United States' Military Sealift Command's Algol class, 30 knot container ships can lift over 23,000 tonnes.[2] Only if equipment is prepositioned can the superior speed of aircraft be exploited to move personnel to be reunited with it. Otherwise aircraft have to make so many journeys that air movement is not as rapid as it might at first seem. As Joel Sokolsky has pointed out in the context of US airlifted reinforcements in a European war, '. . . without prepositioning, it would take nearly 70 days to move five mechanised divisions to Europe after initiation of mobilisation. Sealifted reinforcements, with equipment, could begin arriving in European ports after fourteen days.' Moreover, 'without prepositioning, additional airlift capacity hardly improves delivery times because of the difficulty of organising an airlift of equipment as compared to sealift.'[3]

Sea transport's efficiency in moving heavy and bulky materials extends also to supplies and the long logistical tail required by modern armed forces. No military operation can be sustained for long without large amounts of supplies, and modern armies are more voracious than ever in their consumption. In these circumstances there is often no alternative to seaborne transport over long distances. As the Chairman of the Joint Chiefs of Staff has put it: 'In any major overseas deployment sealift will deliver about 95 per cent of all dry cargo and 99 per cent of all petroleum products.'[4] In 1966 the Vietnam War required 20 million 'measurement tons' of sealift per annum to sustain operations ashore.[5]

For countries like the United States the key word is 'overseas'. Geography is, of course, crucial to the military importance of the sea. If a state is surrounded by land frontiers and has no claims to great power privileges or responsibilities, then the sea is of little or no importance to its military posture. The opposite is the case when a state is an island, of whatever size. Not for nothing did Mahan put geographical position at the head of his list of 'conditions affecting the sea power of nations'.[6] The United States, as a semi-island continent, must go to sea if she is to exert her military power against states beyond Canada and Mexico. For the Soviet Union the situation is more complex. Surrounded by what she perceives to be threats beyond her land frontiers her fundamental strategic requirements are land and air forces. Indeed, if she wishes to project power, air forces and airborne formations are of greater relative importance. It is no coincidence that Soviet airborne forces outnumber similar American forces by about 1.5:1 in men and 3.5:1 in divisions, and American Marines outnumber Soviet naval infantry by 11:1 in personnel and 3:1 in divisions.[7] Soviet airborne forces also tend to have more specialized heavy equipment than their American counterparts. On the other hand, US amphibious capabilities, both as regards specialized shipping and forces for landing, are unmatched anywhere else. Nevertheless, as Admiral Gorshkov repeatedly pointed out, if the Soviet Union is to exert *world-wide* military power there is no alternative than to use the seas to do it.[8]

Many other states must go to sea to project military power. The United Kingdom, Mahan's archetypal imperial sea power, may have lost many of her world-wide military commitments but she has to cross the sea even to maintain her land and air commitment to the continent of Europe, let alone defend herself from the maritime

The uses of the sea

threat to the North. Other complete 'island' states, like Japan and Australia, must see the sea as being of especial importance as a two-way channel for the projection of military power. Both, like Britain, were threatened with seaborne invasion in World War II. Such nations as these (and their smaller counterparts like the West Indian and South Pacific islands) lie at one extreme of geographical strategic sea dependence. Other states are less open to maritime attack but most have some coastline that gives them both naval opportunities and vulnerabilities.

A coastline is always ambivalent strategically. It can be a source of access both for the owner and for a stronger state's maritime offensive forces. Not for nothing were small islands the first and last colonial possessions. Given the traditional dominance of maritime trade, a remarkable amount of the world's population and economic activity is carried out not far from the sea: more than two-thirds of the world's inhabitants live within 300 km of the coast.[9] These people are potentially vulnerable to maritime military power, i.e. sea power, of some kind.

Not only does the sea provide simple access, it also does it in ways that are militarily advantageous. The very extent of the sea provides opportunities for concealment and surprise. Although the revolution in surveillance that has resulted from the development of space-based platforms is undermining this advantage, it is still remarkably easy for maritime forces to lose themselves to view in the world's oceans. The main requirement is that they do not give away their position by emitting electronically or acoustically.

It is this combination of mobility and surprise that gives amphibious operations much of their continuing validity. As discussed in Chapter 2, amphibious warfare has been consistently written off by military commentators, not least after the Gallipoli fiasco of 1915–16. As with all particular historical contexts, the factors that militated against this operation were to a considerable extent scenario dependent. At the outset,

> there was little choice in method of deployment, owing to the mountainous terrain, the restricted beaches and the generally improvised nature of the opposition. Limited movement of some formations between landing sites was possible and was done, but there could be no shifting internally of the main weight, for most of the rifles ashore were sorely needed to hold the existing ring, the overall tactical picture could only be affected by the entry at a selected point of major reinforcements from outside the theatre. And these were by no means readily available.[10]

Execrable operational leadership in the later landings, and an overall

The military medium

commander insufficiently ruthless and too tolerant to deal properly with such incompetence, made failure certain.[11]

In the inter-war period the US Marine Corps, faced with the need to prepare plans for operations against Japan, examined the lessons of Gallipoli and identified where the failures had occurred in command and control, equipment, communications, fire support, and logistics.[12] Measures were taken to solve the various problems. Doctrine was developed and the Fleet Marine Force established. Thus was the background laid for the Pacific counter-offensive of World War II.

Japan, with her amphibious experiences of the War of 1904–5 (which had stimulated Corbett to argue that there was still a place for amphibious warfare in the strategic repertoire), also developed amphibious warfare capabilities and doctrines in the inter-war period which allowed her to take her Great East Asia Co-Prosperity Sphere in the space of a few weeks in 1941–2.[13] Even Britain, understandably the most affected by the Gallipoli fiasco, did not neglect 'combined operations' completely. As one study has concluded: 'the fact is that there existed a doctrine, known only to a handful but nevertheless something on paper, a training and development centre with insufficient personnel and funds but in-being, and a sub-committee of government to put existing doctrine into effect when needed.'[14]

Thus were the foundations laid for amphibious operations that began with the commando raids of 1941 and culminated with the reoccupation of Western Europe. In the latter, air power was a vital component of the amphibious offensive. As Admiral Schofield has put it with reference to the Normandy landings of June 1944,

> the success or failure of the operation depended largely on the rate of build up of the Allied force in the bridgehead exceeding that of the enemy. The Germans possessed interior lines of communication in the form of a network of roads and railways which, unless interdicted, would undoubtedly give them a great advantage in this respect over the sea-borne lines of communication on which the Allies had to depend.[15]

The victory of the Allied air forces, the destruction of the Luftwaffe and the dropping of 76,200 tons of bombs on 80 road and rail targets successfully neutralized much of the Wehrmacht's land power and allowed the Allies to use the sea to put ashore a huge army and its supplies. Between June 6 and June 15, 1944, half a million men and 77,000 vehicles were landed.[16]

If such operations were practical, small-scale operations supported by integral air forces were even more so. World War II was for the United States, and even for the United Kingdom, essentially an exercise in the projection of military power by sea, it might even be argued essentially a 'maritime strategy'. Captain Stephen Roskill, the official British naval historian of the war, tended to argue thus: 'Our total expulsion from the European continent in 1940 forced us to change from a continental to a predominantly maritime strategy: and the reason for the change was simply that no other means of achieving victory then remained to us.'[17]

Roskill concluded that 'the experiences of the last war seem to suggest that, provided all the basic requirements for success in amphibious enterprises are met, they can bring as great benefits as were derived from them in earlier conflicts – both in the field of minor strategy (through small scale raids) and in that of major strategy (by large overseas expeditions)'. 'However', he went on, 'the hazards of such undertakings remain as high as ever.'[18] These have been well summarized by two American analysts. 'Amphibious assault is perhaps the most intricate and difficult of all tactical operations. Preparation for it requires both the creation of peculiar force structures and tactical doctrines, and the development of specialised and often costly weapons and equipment.'[19]

The US Marine Corps has played the leading role in developing these structures and doctrines, its institutional momentum and bureaucratic self-interest helping overcome initial US Army and Air Force, and indeed Navy, doubts about the continued relevance of the amphibious mission post-1945. The Inchon landings of 1950 helped vindicate the amphibious role once more, albeit that this was a relatively small-scale operation carried out by less than one division. Nevertheless, it set the scene for the post-war era, the tendency for amphibious operations to be rather limited and often, unlike Inchon, unopposed.[20]

The British record was similar, but anti-amphibious post-war reaction was more extensive, given the less favourable financial and bureaucratic environment. Despite this, operationally the Suez landings went remarkably well and their lessons were constructively digested afterwards. Using American precedents, and what equipment that could be afforded, a significant British amphibious capability was built up after 1957 for post-colonial policing operations. This was still largely intact in 1982 to retake the Falklands by sea.[21]

The US Marine Corps has used its post-war experience to continue to legitimize a continued investment in amphibious forces that has been massive by the standards of the rest of the world. The officers who man this force have also taken the lead in refining concepts of amphibious operations. The potential of new technologies such as hovercraft landing vessels and tilt rotor aircraft has fostered some

The military medium

radical and interesting thinking on how in the future the sea can be more fully exploited as a military medium. As one young officer put it recently:

> The operational significance of coastal waters has never been fully appreciated. Unlike inland terrain, with its hills, streams, forests, and various other obstacles, the ocean is relatively flat, even in weather conditions that often slow or stop land campaigns, offering amphibious forces a plain on which to conduct initial operations. The advantages offered by this plain can be exploited using new landing tactics based on multiple landing points and rapid shifting of forces. Instead of relatively static and predictable broad landing beaches currently used, much narrower landing points of no more than tens of yards width offer an opportunity to seek out enemy weaknesses. By landing his forces across multiple landing points perhaps in waves of companies, a commander retains the ability to develop situations while committing minimal forces. If successful, initial landing forces can be immediately reinforced by uncommitted units, if not, they can be quickly withdrawn and shifted to more successful landings.[22]

The landing of forces from the sea remains an operation of war with potential in many circumstances. Forces found at sea may be sunk before landing by relatively few weapons. Modern mines make the landings themselves more hazardous. On some occasions, superior land and air communications may be able to bring up a rapid military response that can contain and destroy amphibious incursions. Nevertheless, a proper exploitation of the assailant's advantages makes the result often depend on marginal factors that vary from one case to another. The sea gives much potential for surprise and if the enemy lacks the benefits of a shore transport infrastructure, especially likely on islands or in remote areas, the assailant may actually be better able to support and sustain operations than the defender. Amphibious capabilities, therefore, remain as useful and as challenging as ever.

Amphibious landings, of course, affect primarily littoral areas, as do shore bombardments. The range of bombardment has, of course, greatly increased in the last half century until no significant part of the world's land mass is now beyond the range of naval forces. This is fundamentally a commentary on the capability of sea transport to move bulky objects, like long-range weapons systems, more easily than other means. The combination of first the long-range heavy attack aircraft and the aircraft carrier and then, more importantly, the intermediate and later the intercontinental ballistic missile with the submarine, especially the nuclear-powered submarine, has had a fundamental effect on the world's strategic situation. The condition of 'mutual assured destruction' that provides the strategic underpinning of the modern world is a direct result of the utility of the sea as a medium for the deployment of invulnerable launch platforms.

Once one goes beneath the surface of the sea one enters a murky ultra-complex world where electromagnetic radiations, the main media of both optical and electronic sensors, cannot penetrate easily. Water is a relatively good transmitter of sound waves but even here the uncertainty and unreliability of the sea come more to the aid of the hider than the seeker. The molecules of the water itself absorb sound waves by converting them to heat: the higher the frequency the higher the loss. The sea is also teeming with animal life of all shapes and sizes which both scatter and absorb acoustic radiations and produce a cacophony of sounds of their own. Man-made noises, both ships and drilling operations, add to the noise level. Inorganic matter and bubbles contribute further to the confusion, while the nature and shape of the ocean floor varies enormously, all with serious effect on the acoustic environment.

Although the sea's salinity is usually constant and its pressure increases with depth, it is far from uniform in temperature. There are at least three main temperature 'layers'. The lowest in both depth and temperature is 'isothermic', roughly constant in temperature at between about 3° and 5°C. Coming up at well over 1,000 metres below the surface one then enters the 'thermocline' layer, where the water increases in temperature over 1,000 metres or so from about 5°C to an average of about 20°C. Then there is a surface layer up to a few hundred metres' depth which is again more or less isothermic. Of course, actual surface temperatures vary with time of day, latitude and season from below zero to above 30°C, depending on the area of the world. The boundaries between these layers create an unpredictable acoustic environment, sometimes helping sound propagation along ducts between layers, sometimes blocking off one layer from another. Sound can penetrate from one layer to the next if it is travelling steeply enough but, as might be expected, it suffers both attenuation and refraction. Once in the lowest layer, or deep sound channel, however, sound may go easily for thousands of miles.[23]

As well as these basic horizontal boundaries, which, it must again be emphasized, vary from ocean to ocean, one has vertical discontinuities caused by currents, eddies, fronts, and underwater mountains and valleys. The effect is to create huge shadows in which to disappear from acoustic 'view'. The interaction of the temperature layers and the increase in pressure with depth creates an environment where sound moves up and down in a series of skipping motions.

The military medium

The deflection of sound in the thermocline can both shield submarine platforms suitably placed and allow their detection at longer ranges as the sound channel rises to the surface duct once more, having been deflected upwards by the cool, higher pressure water of the deep isothermic layer. Sound tends to move down to the deep sound channel and then upwards again, close to the surface, at distances of about 50 km from the source, so creating 'convergence zones' about 5 km wide at distances of 50, 100 and 150 km.

With sound at the best of times travelling in these jagged paths and the manifold barriers to detection provided by the discontinuities and background noises of the sea, it is hardly surprising that once a submersible platform loses itself in the oceans it is often hard to find. Despite all these difficulties, acoustics will remain the basis of finding objects beneath the ocean, although future detection systems will also probably be developed to exploit submarine-generated electromagnetic effects. A submarine both creates a 'magnetohydrodynamic' effect by displacing water molecules across the earth's magnetic field and a galvanic current resulting from the electrical potential between the submarine's brass propeller and ferrous hull. These extremely low frequency (ELF) effects can be detected using magnetometers. Current resulting from 'underwater electrical potential' (UEP) is detectable over the longest ranges (measured in miles); magnetohydrodynamic extremely low frequency effects are probably the more persistent as they interact with physical hydrodynamic displacement effects, notably 'internal waves'. The latter can also cause ripple effects on the surface detectable by synthetic aperture radar. Submarines can also cause detectable 'humps' on the surface by physical displacement, temperature variations, and even 'bioluminescence' among marine life. Leaving aside the question of active decoying, all of a submarine's effects are also produced by natural means, winds, natural currents, whales, etc. Long-range, high certainty area submarine surveillance is thus still a long way off, even if supplementary non-acoustic forms of tactical and operational detection are becoming available. UEP mines are reportedly already deployed by the Soviet Union. Magnetic anomaly detection (MAD) of the distortions caused by a submarine in the earth's magnetic field has been useful for short-range ASW since the World War II years, and could be developed further.[24]

The key to all this is that if the sea was a uniform static medium it would be relatively easy to detect a number of indicators that objects were present. However, it is neither uniform nor static. Only the most sophisticated computer analysis allows the natural phenomenon to be distinguished from the potential target and most, if not all, non-acoustic indicators produce insufficient information

The uses of the sea

Upon which computers can even base their analysis. Sonar applied to modern signal processing techniques does provide some transparency to the seas, but it is a variable and uncertain transparency even now. There is also always the possibility of deliberate deception. The sea will for long remain the best place in which to hide and will thus remain the world's most important medium for the deployment of those forces upon whose invulnerability strategic stability rests.

It is not, however, just the physical characteristics of the oceans that make them very attractive as a military medium, both in normal times and in crisis. The interaction of the sea's ease of movement, especially of stealthy movement, with its international status as the 'great common' gives it unique advantages as a medium for the deployment of military power, especially in nuanced situations short of open conflict. Naval power projection forces can be kept out of sight or deliberately displayed. They can be manoeuvred in threatening or non-threatening ways. They can be sent to cruise in international waters in or close to a potential crisis area without raising all the problems of acceptability and commitment that a deployment of land forces or land-based air forces would entail. Naval forces can bring air and missile power to bear without the need to negotiate overflying rights and/or engage in difficult and elaborate airborne refuelling exercises. Usually, proximity to the target also facilitates greater accuracy and discrimination in the delivery of firepower. Forward presence can be maintained for long periods if required. Even amphibious forces can be kept poised at sea for weeks on end, giving options for intervention that can be unveiled or utilized as and when required. If circumstances do not develop in the war feared, perhaps because of the maritime deployment or perhaps despite of it, perhaps for other reasons, those forces can be withdrawn without too much diplomatic penalty. Maritime forces also lend themselves to quick 'in–out' operations, such as those required to evacuate nationals from dangerous situations.

It is in the area of providing contingency forces for decision makers that sea power really lives up to its reputation for silence of operation. In 1967 the British commando carrier (LPH) HMS *Albion* disappeared into the Atlantic for many weeks with her embarked commando, artillery battery and helicopter squadron. This gave Britain the option to take limited action in the opening days of the Nigerian Civil War, perhaps to create a neutral zone if requested, perhaps to act more unilaterally to evacuate civilians. The ship's absence from scheduled exercises was covered by the Arab–Israeli crisis in the Middle East that broke about the same time.[25]

Even more covert by its very nature was the despatch of the

nuclear-powered submarine HMS *Dreadnought* to the vicinity of the Falkland Islands at the end of 1977, to be unveiled should a threatening Argentina be thought to be about to invade. The aim was to do this without 'provocation' and the two frigates sent to provide vital communications links were kept as far away as possible. There is some debate as to whether Argentina was informed of these deployments at some stage, but the main aim seems to have been not to force Buenos Aires to back down but to provide a means of setting up a defensive 'exclusion zone' immediately should the situation warrant it. It is interesting to speculate what would have happened in 1982 if the Thatcher Government had shown the same foresight as that of its Labour predecessor.[26]

Submarines by their very nature are usually less useful than surface ships if overt presence is required. Even in the above case the only way HMS *Dreadnought* could have proved her presence would have been by sinking something. They are, however, useful in other covert roles such as intelligence gathering. Both superpowers and others exploit submarines to enter each other's sovereign waters, to gather information on each other's sensitive installations and assets on, above, and below the waves, and to plug into communications cables to monitor sensitive traffic. Incidents often occur, such as collisions and even submarines running aground, as happened in the Soviet 'Whiskey on the rocks' incident in Sweden. Except in such blatant cases, however, things can easily be kept quiet to both sides' benefit.[27]

Given the extent of the high seas, however, surface vessels have a key role to play in intelligence gathering and analysis. The platforms used include warships, merchant ships and specialized auxiliaries (AGIs). *Jane's Fighting Ships* lists no less than sixty-three specialized intelligence collection vessels in Soviet service. Most are based on trawler hulls but there are also vessels in the 3,000–4,000 ton class.[28] The United States has no ships actually designated 'AGI' but its use of specialized surveillance vessels, like the ill-fated *Liberty* and *Pueblo*, is well known. Perhaps America's most notable current AGI is the 'missile range instrumentation ship' *Observation Island* that monitors Soviet missile tests and electronic emissions in the Northern Pacific.[29] Surveillance, especially by surface ships, can be an important tool of crisis management, having the double effect of demonstrating an interest in the area in question and the forces under surveillance, and providing an accurate monitor of the other side's activities. The actions of surveillance platforms can also be used to show resolve and exert pressure.

Those states likely to be the victims of these uses of maritime power may resent the intrusion, especially if they are unable to take appropriate and effective countermeasures. The 1986 UN Study, 'The Naval Arms Race', articulated concerns that are widely held in the Third World.

> When employed on normal deployments as part of national peacetime tasks, activities by world-wide and blue-water navies outside their own territorial and regional areas can become a significant political factor in regional and local situations. . . . Many regional states may become concerned at the implications of such deployments for regional security and strongly dislike the implied threat, real or perceived, of external intervention in the regional or internal affairs of other states.[30]

It called for 'the exercise of moderation and restraint on the part of all, . . . if further threats to security are to be contained'.[31]

Happily, the nature of maritime forces, working in a neutral medium under tight national control (a control that is becoming tighter all the time), lends itself to the exercise of moderation and restraint. Although the very presence of warships may act as a provocation to an opponent or a too readily available resort to force, more usually the actors in a particular crisis exploit the opportunities for limitation and discrimination in response. Tight 'rules of engagement' (ROEs) are usually laid down to regulate the way weapons can be used and platforms manoeuvred, and the circumstances in which actions can be carried out or threatened. They are especially important at sea, where there are not the clear demarcations provided ashore by national boundaries. ROEs can be manipulated and sometimes ignored by commanders on the spot, as when in the Falklands War the British commander tried to unleash his submarine in contact with the cruiser *General Belgrano*. The signal was intercepted and countermanded, but the unauthorized action became an exercise in effective manipulation as it demonstrated all too effectively to London the gravity of the situation. The ROEs were, therefore, changed to allow attacks on all Argentine warships outside territorial waters.[32] The natural dynamics of naval confrontation can sometimes force escalation despite attempts by the actors in the crisis to keep the lid on the situation; when ships are sunk the symbolic sense of loss can be great and the casualties can be heavy.

These escalatory pressures must be borne in mind in analysing ideas for limited conflicts at sea between major powers. There are some precedents, such as the 'Phoney' period of World War II over the winter of 1939-40. This period of restraint, advised by the conjoint fear of the air raid and renewed trench warfare, seemed of particular relevance when the nuclear age made the price of total war

The military medium

unacceptable. A conflict limited to the oceans might, in the words of Professor Martin, be 'more likely to remain within the bounds of a regulated policy contest than one that wrought destruction on homelands and civilians even only as the collateral effect of a would-be intra-military contest'.[33] During the Berlin crisis of 1959–61, the United States studied various graduated responses against Soviet, DDR or other Warsaw Treaty Organization (WTO) shipping 'which would be preferable to directly challenging the Soviets along the NATO frontiers because it carried less risk of escalation to nuclear war'.[34] By 1967 there was an Atlantic 'War at Sea Plan' which was 'specifically designed to apply a measure of force in support of a national policy of determined persuasion, or alternatively may be implemented as a controlled response to Soviet military aggression short of strategic nuclear attack'. Using well-defined ROEs the US Navy would have a number of options 'to employ military action against the Soviet Union under circumstances in which implementation of the general war plan is neither a rational nor credible threat'.[35]

This kind of planning exploited the very 'secondary strategic value of the sea to the Soviets. If Russia did not depend heavily on the seas to support its economy or its military power in Central Europe then striking at Soviet vessels would not elicit a massive retaliation by the Soviets.'[36] Since the late 1960s Soviet dependence on the sea has increased further, but it remains less than the West's. This might, however, be held to advise against the West, which has more at stake, deliberately exposing a vulnerable flank to Soviet pressure. Turning the above argument on its head it is sometimes argued that the Soviets themselves might begin a maritime offensive against Western shipping. The graduated nature of the pressure might allow time for negotiation. However, a number of problems immediately present themselves to such a plan. The international nature of modern shipping would force the attacker to take on many countries, not just NATO. Given the use made by the West of land-based sensor arrays and aircraft, it would be exceptionally difficult for land operations to be dispensed with entirely. Moreover, the balance of maritime forces is such that Western maritime power might be projected in several ways that the USSR would find unwelcome, especially against its more far-flung bases at home and overseas. Even if the maritime inderdiction campaign was a success, an increasingly desperate West might well consider such 'horizontal escalation', perhaps even increasing the intensity of conflict; the results of a successful blockade can be anything but 'limited' in principle or practice.

Despite all its apparent advantages as a neutral *place d'armes*, one

The uses of the sea

must be careful not to overstate the utility of the sea as a self-sufficient military medium. It is the very interaction of the sea and the land, where, as Corbett pointed out, people actually live, that gives sea power its fundamental strategic salience both now and in the future.[37] Maritime coercive power is exerted in a remarkable number of different ways and from an equally large variety of mobile platforms: a very wide range that helps give sea power its inherent flexibility and controllability, both now and in the future. As we shall see in Part III, every contingency is covered from mild constabulary pressure to all-out nuclear holocaust.

Part III
The changing shape of naval war

Chapter 5

Weapon systems

This part of the book addresses itself entirely to the current state of naval technology and the paths of its likely future development. First, we look at the weapons, together with the sensor, information, and control systems that are the keys to their effectiveness. As weapons systems are means to ends not ends in themselves it is probably best to examine them in the context of their functions in the various dimensions of naval operations: anti-surface ship warfare (ASUW), anti-submarine warfare (ASW), anti-air warfare (AAW), and operations against the shore. The special question of nuclear weapons will be dealt with last.

Anti-surface warfare

Anti-surface warfare is in a sense the easiest of naval tasks, given the vulnerability of the target to damage above and under the water. A range of technologies are available for this role: guns, missiles, and torpedoes. From the middle of the sixteenth century to the beginning of the twentieth, guns were the most important instruments of bringing force to bear against an enemy's surface ships – although for a surprisingly long period they were more a means of a 'softening up' for a human assault or obtaining a surrender than a very efficient means of exerting decisive damage on the enemy's fabric or watertight integrity. Only with the advent of the long-range shell firing gun in the middle of the nineteenth century did artillery become capable of decisive results on its own. But its reign was soon challenged by underwater weapons such as effective mines and torpedoes, which were born at about the same time. Nevertheless, guns retained a considerable range advantage and, directed with increasing accuracy by some of the world's pioneer analogue computers, remained the dominant naval weapon until well into the twentieth century. Only in the early 1940s did they have to yield supremacy to air-delivered bombs and torpedoes as the most effective way of neutralizing an enemy's major naval assets. They had, however, been replaced by the torpedo as the main means of sinking merchantmen a quarter of a century before.

One does not, however, necessarily always want to sink an opponent, especially in conditions short of all-out conflict. At the time of the Cuban missile crisis in 1962 it dawned on the US Navy that their latest all missile Leahy class frigates (now classified as cruisers) could not fire a shot across the bows of Soviet merchantmen: 76 mm (3 inch) guns were soon added, and the design of the follow-on Belknap class was altered to include a 127 mm (5 inch) weapon. At about the same time the Royal Navy was finding that 114 mm (4.5 inch) guns were a somewhat inappropriate answer to small Indonesian sampans; old 20 mm guns of World War II vintage soon reappeared. These lessons retain their relevance. A weapon with a relatively rudimentary fire control that can deliberately miss the target as well as inflict precise and limited damage upon it will remain a warship's minimum necessary armament. Tube artillery in the 20 mm to 76 mm range would seem the obvious answer.

Guns, however, will remain useful for more demanding ASUW roles also. The combination of rocket assistance and simple infrared/laser homing technology is giving artillery the characteristics of range and accuracy of the shorter ranged anti-ship missiles. Given the capacity of a ship to carry a much larger number of rounds of gun ammunition than relatively large and bulky missiles, this would seem to give guns advantages in ship-to-ship engagements of medium range, e.g. out to 30–40 km. Not only can more targets be engaged but the capacity to fire a number of projectiles at each individual target increases the chances of a kill. It will be more difficult to give 'cannon-launched guided projectiles' (CLGP) the autonomy of missiles given the limited payload available for guidance systems and the robustness required to withstand the shocks of tube launch. Near-term CLGP will also require some external source of designation, either from a helicopter or RPV, or possibly from the laser range finders now being increasingly fitted to surface ships themselves.[1]

To give CLGP an adequate payload, a relatively large calibre is probably essential. This will probably cause future gun calibres to edge up slightly from 114–127 mm (4.5–5 inch) to 155–203 mm (6–8 inch). Anything bigger, however, will cause the gun to dominate the design of a ship to an undesirable and unnecessary degree, given the need for a future warship to be capable of operating against underwater and air targets as well as against other ships. Very heavy calibre artillery, as mounted in the superannuated American battleships, could notionally deliver large and powerful CLGP to 80 kilometres or so. New and relatively expensive ships built primarily

Weapon systems

around super-heavy artillery would probably be over-specialized and not cost-effective, given available ASUW alternatives. It is possible, however, that, as proposed in the United States, new gun-armed ships with up to 203 mm (8 inch) main armament will be built, but the primary purpose of such vessels will be anti-shore warfare rather than anti-ship.

Until the Falklands War gave a dramatic demonstration of their capabilities, torpedoes were falling out of fashion as weapons of ASUW. This was partly because ships had rarely been targets in post-1945 naval actions. Also, torpedoes are highly lethal: they cannot be used except actually to sink the enemy. In the three and a half decades after 1945 only in the Indo-Pakistan War of 1971 were torpedoes operationally used in conflict by a Pakistani submarine to sink an Indian frigate. Then in 1982 came the dramatic sinking of the Argentine cruiser *General Belgrano* by the British submarine *Conqueror*, and the less successful attacks by the Argentine submarine *San Luis* on British surface combatants. The weapons used against the *Belgrano* were classic indeed: Mark 8s, half a century old in basic design, although perhaps the best torpedoes of their kind. Powered in this case by hot gas, or sometimes by electricity, such weapons have their course set before launch and are best fired in salvoes. (*Conqueror* fired three of which two hit.) The effect of two 340 kg warheads exploding below the waterline of even a well-built surface ship like a World War II cruiser was devastating, but these classic torpedoes have to be released close to the target. *Conqueror* was only 1,400 yards away. Achieving this is difficult against an enemy who is capable in ASW techniques, or indeed one who is manoeuvring evasively. This is not to say that such torpedoes will not continue to find a place in naval operations, but their usefulness will be limited to unescorted and unarmed shipping and/or vessels of lesser states.[2]

Long-range capability of an indeterminate kind can be obtained by pattern running, i.e. causing the torpedo to steer a zig-zag course after being fired into a group of ships. This might still have some utility. It is better, however, to fit the torpedo with some acoustic homing device to guide it on to its prey. Wire guidance from the launching platform can be used to guide the torpedo to the vicinity of the target, as revealed by the launching platform's sensors, e.g. a submarine's sonar. When the torpedo is safely locked on to the target's noises it can be left to run its course. This can have its problems: as the Argentinians found when the guidance wires of their German torpedoes snapped.[3] Wire guidance will still be used in many future ASUW torpedoes and in common torpedoes for ASW and ASUW use, but the latter may not be the most economic

The changing shape of naval war

solution. Surface targets are relatively noisy, and acoustic homing torpedoes often only need to be fired into the target's area seeking passively to find them. Whether wire guidance is used or not, ASUW torpedoes can be much less sophisticated than their ASW brethren, and the US Navy plans to procure a low-cost acoustic torpedo (LCAT) for ASUW duty at a cost of only 20 per cent that of its fully capable ASW cousin.[4]

Torpedoes are, *par excellence*, the weapon of the submarine, but small surface craft have traditionally been used as delivery means also. This continues with, for example, tubes for wire-guided torpedoes being fitted to supplement the missiles of West German fast-attack craft. Sweden deploys large wire guided ASUW torpedoes in shore batteries. The latter are some of the few torpedoes to use hydrogen peroxide (HTP) propulsion. Batteries are the more common modern propulsion systems for ASUW weapons, although modern monopropellent heat engines are making an appearance and will become more common in the future.[5]

Anti-surface torpedoes are traditionally 21 inch (533 mm) weapons, although smaller torpedoes have been deployed, primarily as anti-escort self-defence systems in submarines. In World War II, the Japanese successfully used long-range 24 inch weapons to give ranges comparable to gunfire, and such large torpedoes may be making something of a comeback in an interesting novel form. It is reported that the Soviet Union is now deploying large Type 65 torpedoes of 650 mm calibre (26 inch) with advanced closed-cycle (probably HTP powered) propulsion systems. Their reported range is 100 km at 30 knots or 50 km at 50 knots, using 'wake-homing' techniques by which sensors look for the turbulence in the wake of a target and then home in upon its source. This is probably done acoustically with a small torpedo-mounted sonar sensing the turbulent water in the ship's wake. Other more advanced types of wake-homing devices are conceivable, e.g. 'a passive optical or pressure/temperature sensitive instrument or one which measures variations in sea-water electrical conductivity associated with turbulence'.[6] Long-range wake-homers can be fired into the wake of an enemy's surface force at some considerable range. They then 'look for' indications of wakes and by means of a series of diminishing zig-zag paths direct themselves towards the source of the effect. Such weapons concentrate on the vulnerable stern areas of a ship, where the propulsion and steering gear of the vessel is concentrated. They might not sink a very large enemy ship such as a carrier but they would be almost certain to put it out of action. A torpedo of this size might carry a rather large warhead (400 kg) with a large kill radius. It is interesting that the introduction of such weapons has coincided in the Soviet Navy with the end of the production of submarines with shorter-range anti-surface ship missiles; the role has apparently been taken over by SSNs that are equipped with the new torpedo tubes. The effect of a hit of a single long-range 'wake-homing' torpedo on a US carrier would be potentially catastrophic: much more so than even a couple of missile hits.

More effort, therefore, may have to be given to torpedo countermeasures. Major navies already have acoustic decoys, like the US 'Nixie' or the British Type 182, that can be streamed behind a ship to divert homing torpedoes. These systems are already inadequate against later torpedoes. Under development are more advanced reactive decoy systems, such as the British 'Expendable Intelligent Decoy', that can be dropped into the water either to use noise jamming to distract passive homers or to seduce active homing torpedoes by retransmitting the torpedoes' transmissions in such a form that the enemy weapon homes in on the wrong spot. The French are exploring a reactive expendable decoy system to be applied first to their new carrier. 'Pyro-acoustic' countermeasures are being used, pyrotechnics to create noise or gas generators to create bubbles to 'mask' the ship. Such systems may help deal with modern acoustic homers, although the advent of formidable 'wake-homers' may make anti-torpedo torpedoes, as suggested recently by one officer, the only answer. This could be an expensive and complex option.[7] Another potential defence against this underwater threat is the use of short-range ASW rocket launchers as anti-torpedo defences. All large Soviet combatants either carry special RBU-1000 sextuple launchers on each quarter for this duty or, in the case of the 'Udaloy' class ASW cruiser, mount their longer range RBU-6000 weapons, so they can apparently double in this role. The Soviets deny that this is so but similar systems might have some anti-torpedo potential in the future.

The most currently fashionable method of attacking surface vessels is to use anti-ship missiles. This form of naval warfare was pioneered by the Germans and Japanese in World War II, the latter using manned guidance. It can be effective, as demonstrated by the Argentines in 1982 when Exocet missiles were used to sink an important supply ship and a destroyer. The Soviet Union has developed a range of anti-ship missiles for deployment on a spectrum of platforms: airborne, submarine, and surface. The smallest of these have scored successes in the Middle East and Indo-Pakistan conflicts. In general, however, Soviet missiles, and the Chinese copies of them, have tended to be relatively large and fly relatively high.[8] Many current longer ranged Soviet missiles are also relatively slow, giving a requirement for mid-course corrections and the launching platform, if a submarine, having to stay in a vulnerable surfaced position as it guides its weapons to the target before they lock on. This drawback is being overcome by newer, more autonomous missiles like the SS-N-19, which is an advanced weapon capable, if necessary, of submerged launch and with a range of some 450 km at supersonic speed. Once fired at a target provided by an external source, e.g. satellite or aircraft, the missile can get quickly to the point where its radar head can acquire and home in on the target.[9]

Some Soviet missiles already at sea have the skimming characteristics of modern Western missiles such as Exocet. This rocket-propelled weapon is an air-launched missile with a range of about 60 km and a speed of just under Mach 1, giving a flight time of about two and a half minutes. The launching aircraft, usually a small fighter-type machine, exposes itself very briefly to radar detection to programme the weapon's autopilot. The missile then skims over the sea at a height of a few metres (dependent on sea state) turning its radar homing head about 30 seconds from impact to give the target the minimum chance of taking countermeasures.

Exocet, which has surface and sub-surface launched variants, is in the middle range of anti-ship missiles. Jet propulsion gives longer range, as with the American Harpoon and British Sea Eagle, although problems of target acquisition after a long flight act as a potent constraint on effective range as opposed to theoretical maximum. This is an especially serious constraint for very long range anti-ship missiles like the American Tomahawk. The latter's ability to find a target covering its advertised range of 400 km at Mach 0.7 has been doubted by leading analysts. Data links to update the guidance of missiles for an external surveillance source may be necessary for such weapons to exploit this kind of range bracket.[10]

The new ANS (*Anti Navires Supersonique*) being developed co-operatively by France and Germany will use ramjet propulsion to enable it to cover its range of approximately 175 km at no less than Mach 2 at sea level (these speeds have previously only been achieved by Soviet AS4 and AS6 air-launched missiles, both of which have a high altitude mission profile). ANS will be an 'intelligent' missile, i.e. fitted with sufficient on-board computer capability to enable it more fully to cope with enemy movements and countermeasures, and even if the latter are deployed, weapons of its generation may well be able to think their way through many of the more obvious techniques. The weapon will be easily reprogrammable as the measure/countermeasure dialectic evolves.[11]

Aircraft are probably going to remain the most effective anti-ship missile launchers as they are best able to find and identify targets as well as only having to remain for the most limited period in the latter's vicinity. Modern integrated surveillance techniques using satellites and reconnaissance aircraft, however, might well provide missiles from surface or even sub-surface platforms with sufficiently good initial targeting information for long-range engagements. The introduction of maritime surveillance aircraft fitted with inverse synthetic aperture radar (ISAR), with its ability to give a high-definition picture to any receiving platform, allows long-range missiles, such as Tomahawk, to be fully exploited by surface ships over ranges of hundreds of kilometres. Modern towed array sonars provide surface ships or submarines with autonomous very long range initial detection and surveillance capabilities. These do not always work (e.g. HMS *Splendid*'s inability to find the Argentines' carrier battle group in the Falklands War), but they do greatly extend the detection horizon of the platform on or in the sea and, if the weapon carried has a sufficiently wide search pattern, allow an extended attack horizon. Helicopters or drones can be used to give supplementary surveillance and targeting information to surface ships. The key to all this, however, remains discriminating real targets from what will probably be a jumble of contacts.[12]

Increased missile speed will also make long-range engagements much easier. An alternative approach to high-speed cruise missiles might be the use of terminally homing ballistic missiles. For a time in the 1970s the Soviets seemed to be developing such a weapon specifically for use against ships, the SS-NX-13. A two-stage rocket launched a manoeuvring post boost vehicle to an altitude of 250 km where its radar sensors searched for the target. Having picked it up and locked on, the PBV reoriented itself and using two rapid rocket burns propelled itself in the direction of the target at very high speed. The actual re-entry vehicle was then released just above the atmosphere to fall on the target. The system seems to have depended on the use of a nuclear warhead but modern manoeuvring re-entry vehicle technology could allow it to home in all the way. The inevitable difficulties of development, coupled with SALT limits on ballistic missile launchers, caused the abandonment of this system, but it possibly could be resurrected to formidable effect. The range was between 150 and 1,000 km.[13]

Over short ranges modern, semi-active, radar-homing surface-to-air missiles now have quite good secondary surface-to-surface capabilities over ranges of about 35 km or so, out to the horizon. The main problem is how to decide on the relative priorities for these valuable weapons, given limited ammunition loads. At the lowest end of the scale, specialized anti-ship missiles of around 20 km range will still be useful for use from fast and manoeuvrable platforms, such as fast-attack vessels and helicopters. The Israelis demonstrated the utility of such missiles when their fast-attack craft equipped with Gabriel missiles worsted the SS-N-2 equipped opposite numbers of Egypt and Syria in 1973. The latter's missiles were more suitable for engaging larger warships than other fast-attack craft. Other advanced Western short-range missiles, e.g. the Norwegian Penguin, will continue to be effective in relatively enclosed waters. Helicopters also can usefully be fitted with such devices to defend their operating platforms from attack by smaller craft, and to expand the ASUW coverage of a surface ship. A missile designed specifically for this duty is the semi-active homing Sea Skua used in the Falklands to damage an Argentine patrol vessel.

The limited lethality of this weapon illustrated a general problem with all missiles. A hit above the water line is much less catastrophic than one below it, unless the target has special vulnerabilities. The Exocets that scored successes in the Falklands seem to have been incendiary devices as much as anything else, their rockets setting fire to vulnerable wiring systems or cargo. The same seems to be true of hits by the Exocet missiles in the Gulf War. Indeed, there is a persistent doubt about the reliability and effectiveness of Exocet's warhead. Only one of the pair that hit USS *Stark* exploded. Even if it does explode, glancing blows, e.g. as suffered by HMS *Glamorgan* in the Falklands, do not cause catastrophic damage. Measures are being taken to provide ships with lightweight armour to protect against warhead damage, and to harden their electrical systems against the incendiary effect of missile hits. These will be countered by improvements in both warhead lethality and reliability, but medium-sized combatants may gradually become significantly less vulnerable to anti-ship missiles. As the Gulf War demonstrated, unless they are hit in flammable cargo or storage areas, large merchantmen posses considerable resistance to such weapons; warships are, paradoxically perhaps, much more vulnerable although again the bigger the ship the better in damage resistance terms. Smaller vessels, like the Libyan fast-attack craft, and the unknown mystery ship sunk by the US Navy's Harpoon missiles in the incident of March 1986, will remain the most lucrative missile targets.

As steps are taken to diminish the vulnerability of targets so the benefits of combining missiles with final *underwater* attack might seem attractive. Such systems were indeed considered late in World War II when armoured targets were still a significant problem. The British developed a large unguided rocket called Uncle Tom to strike underwater, and various more exotic air-launched torpedo devices were worked upon. The relative lack of toughness of most post-war naval targets, and the priority for ASW and the advent of nuclear weapons, sounded the death knell of these advanced systems, but the Americans did deploy briefly in the mid-1950s an interesting system called Petrel, an air-launched, acoustic homing torpedo that was launched by a maritime patrol aircraft five miles from its surface target. A jet engine and radio command guidance system placed the torpedo in the correct spot for it to complete its run under water. Technical progress over the last thirty years might make a more modern variant of such a device a truly formidable weapon.[14]

Moving to underwater attack would also help deal with the classic forms of countering missile attack, i.e. covering the target area with a chaff cloud. The chaff prevents the incoming missile locking on to the target ship. This technique was used with some success by the Israelis in 1973, and also by the British in the Falklands War, although as the sinking of the *Atlantic Conveyor* demonstrated, missiles can be potentially diverted from escorts on to higher-value shipping in their care. Chaff equipment is being made much more effective. The current Barricade rocket equipment used by NATO navies is controlled by a computer system. It distracts by putting up decoys before the missile acquires the real target, or confuses by placing additional ship-like echoes on the missile's acquisition radar, thus delaying or preventing target selection. If a persistent missile does manage to lock on, the chaff cloud can be used together with jamming to 'steal' the target from the missile, or the chaff can be put in the right place to seduce the missile away from its target. The system computer generates an optimum course for the ship to steer for the missile to break lock. Even more advanced active countermeasures are becoming available, e.g. Marconi's Siren system, which sends back to an incoming missile a precisely tuned radar 'song' to draw it away from target. Only one rocket is required per threat which helps cope with multiple attacks. Floating decoys have also been developed. For example, Marconi produce simple 'towed offboard active decoys' for use with any kind of ship. Developed for the Gulf, such TOADs can be 'tuned' to any likely threat. The little boat-like decoy searches for missiles automatically. If it detects a threat it sends jamming signals to seduce the missile from the target vessel.[15]

This is not to say that missiles will be denied effectiveness in the future. They will remain most potent threats, especially when deployed in large numbers to swamp defences. Not all targets will be equipped with modern countermeasures and a large number of especially vulnerable targets will be at risk. Nevertheless, it seems unlikely that the anti-ship missile will drive shipping from the surface of the sea. Like the torpedo, which made a somewhat similar entry onto the naval stage a century or so ago, the missile's capabilities have been overrated. No single 'wonder weapon' can affect something as complex as maritime operations. In certain circumstances it might prove decisive, in others ineffective, especially as countermeasures develop.

A good example of the complexity of the factors governing the effectiveness of naval weapons systems is the effectiveness of 'iron' bombs dropped by small fighter aircraft. Given the strength of modern ship defences, it is not easy for an aircraft to use an ordinary bomb against a well-protected naval target. As the Argentines found in 1982, even if hits are scored the delivery altitude required is usually so low that the weapon often fails to explode. If the defences are weak, however, the attacker's delivery altitude can be higher and these weapons can be all too effective. Retarded bombs help solve the problem of ultra-low-level delivery and these may well be a significant threat to ships, especially in coastal waters, for some time to come. The fate of HMS *Ardent* in 1982 was a potent demonstration of this.[16] Cluster bombs, as used against Libyan patrol craft in 1986, also have advantages, especially against relatively small ships, while laser-guided bombs homing on the reflected energy of a laser designator proved formidable anti-ship weapons in the Gulf in 1988.

A final form of anti-surface warfare to be considered is an old and, until recent incidents in the Gulf, often a neglected one: mining. Still perhaps the most effective means of closing a harbour in war or violent peace is the underwater explosive device set to be exploded by a passing ship. Contact mines are still available, but 'influence' mines have greater potential using the three main signatures of a ship: magnetic, water pressure, and acoustic. More than one sensor can be combined in the same weapon and ship counters can be fitted to delay action to a later target than the first to pass over the weapon. Modern electronics allow mines to be given a remarkable degree of 'intelligence' to deal with the information provided from their various sensors. Mines can be programmed to respond to specific signatures, e.g. large, lucrative ship targets rather than small, mine countermeasures (MCM) vessels.

The classic mine consists of a large explosive charge placed on the bottom or moored to it. These are only useful if exploded relatively close to the target and are only effective, therefore, in shallow water. The last quarter of a century or so has seen the development of 'mobile' mines which can be laid in deeper water. More primitive kinds 'oscillate', i.e. move up and down on a pattern searching for targets. Others 'home' on to a target either themselves or by releasing a 'homing' torpedo. Such mines tend to be used primarily against submarines, but they have a wider utility and will probably be more widely used against surface vessels too. They also create a set of new hazards for mine countermeasure forces, which will have to worry increasingly about being 'outranged'.

Mines are the least appreciated of naval weapons. Their use is often seen as an unwelcome and rather passive and 'defensive' alternative to more 'normal' types of naval warfare. There is perhaps some truth in this, and the Russian propensity for mine warfare has much to do with their traditional lack of expertise in more conventional naval operations. This, however, should not lead one to discount the effectiveness of mine warfare. It has been very effective in the past (e.g. in the opening weeks of World War II in British coastal waters, in the final US blockade of Japan, and in closing Haiphong in 1972). It could be even more effective in the future. Mines can be laid by a wide variety of platforms: aircraft and submarines as well as surface ships of all shapes and sizes. Covert use by unofficial groups as well as by governments has taken place and may well occur again.

Mine countermeasures thus have a growing importance and there has indeed been a notable increase in interest in the subject among the world's navies. The Royal Navy, for example, which began a large fleet of MCM vessels in the 1950s but then rather lost interest as strategies changed and alternative, more spectacular priorities were preferred, is currently investing over a billion pounds in efforts to improve its MCM potential. The enhanced security of NATO reinforcement shipping is a major aim of this programme (as well as the free passage of RN and USN ballistic missile submarines).

A real problem, however, is that countermeasures against modern mines are far from easy. The easiest and most traditional method of dealing with mines is sweeping, using wire sweeps to cut the moorings of fixed mines causing them to come to the surface for disposal. Influence mines laid on the bottom in shallow waters can be exploded by magnetic or acoustic sweeps simulating passing ships. The effect to which pressure mines react cannot, however, be easily simulated and simple ship counters or more intelligent discriminating techniques can relatively easily defeat sweeping efforts. Unless the sweeping is done by helicopter, the MCM vessel itself is always at risk. One way of overcoming this problem is to use robot minesweepers controlled from a mother ship, as pioneered by the West Germans with their Troika system. The Swedes have also developed remotely-operated acoustic-magnetic sweep vessels called SAMs for use with their Landsort class MCM vessels.

The manifold drawbacks of sweeping in the modern era mean that mine hunting often has to be resorted to, which is always a time-consuming business. The normal technique is to use human divers inspecting and disposing of, or removing, objects discovered by a precise sonar equipment in the minehunter. The latter may be neutralized by adverse water conditions, but variable depth mine-hunting sonars are being developed to deal with this problem. As an alternative to human divers, TV equipped mine-neutralization vehicles can be sent down into clear water of limited depth to inspect objects and, if necessary, lay charges next to the mines. Advanced Remote Minehunting System (ARMS) is a British produced, remotely operated vehicle that does not require a specialized MCM platform. Mine hunting has its own hazards: mines may be booby trapped or programmed, if mobile, to home on to mine-hunting sonar.

Much work has still to be done to improve MCM capabilities (e.g. pressure sweeps) to counter the guile of the offensive user of mines. These unspectacular weapons look like retaining a considerable degree of effectiveness for sea denial for many years to come.[17]

Anti-submarine warfare

ASW has come a long way since the first submersible boats led to such desperate expedients as training seals or seagulls or patrolling harbours with bags to place over submarines' periscopes. Anti-submarine torpedoes, deployed from a variety of sonar-equipped platforms, now hunt down their quarry using their own effective acoustic sensors. These are the major ASW weapons systems and are likely to remain so, but other methods, mines and even traditional depth charges, will still have a significant part to play in certain circumstances.

An anti-submarine torpedo must conform to a set of most demanding characteristics. One expert had identified six criteria for any such weapon:

- It has to be fast enough to attack and if necessary overtake any evading target.
- It requires sufficient range to reach, attack and re-attack if needed.
- Its detection capability must be more than adequate to overcome inaccuracies in placement.
- It must be able to distinguish between real and false targets and be able to handle or disregard countermeasures.
- It must be launchable into and effective in the shallowest depths in which submarines can operate and yet be able to attack them at their maximum depth.
- It must be lethal.[18]

Weapon systems

These requirements sometimes conflict. A torpedo must catch its target without creating so much noise that it alerts it or masks the weapon's own sonar or that of its launching platform. A physically small sensor must be combined with torpedo electronics that can distinguish between the sound of the quarry and other sounds, including the torpedo's own. As the expert quoted above went on to write, torpedo development is a problem-filled exercise:

> A wide variety of techniques are brought into play – mechanical or electrical propulsion, hydrodynamics, warhead design, electrotransducer development, amplifier design, the latest signal processing including correlation techniques, beam steering, digital processing, algorithm definition, computer architecture guidance, radiated noise suppression, advanced power controls, flow noise controls, simulation techniques, analysis and tracking – a formidable list which makes it difficult to succeed without prodigious effort and allocation of large financial and manpower resources.[19]

And not even necessarily then as the British experience, at least until recently, shows.[20]

At a price in both time and effort, formidable weapons can be created. The United States is now deploying, and the United Kingdom developing, submarine launched 533 mm (21 inch) torpedoes which combine high energy 'monofuel' propellants (stable forms of nitroglycerine), external combustion, reciprocating gas engines or gas turbines and water jet propulsion to propel them quietly out to 30–40 km at speeds of about 50–60 knots. These torpedoes, the American Mark 48 and British Spearfish, are highly intelligent devices with on-board computers which can adapt to the complex changing environment in which the torpedo finds itself. Guidance wires allow the torpedo to interact with the parent submarine and its more powerful sonar and computer capability. The weapons are fully capable of attacking autonomously, however, if the wire was accidentally broken or when the launching submarine is sure of 'lock on'. At a million dollars per unit, however, these weapons are expensive and most navies will have to rely on similar weapons of reduced capability like the current propeller driven, battery electric powered British Mark 24, with only half the range and half to two thirds the speed of the Mark 48.[21]

If further increases in performance are required then new propulsion technology will be necessary. Open exhaust engines like those used in the Mark 48 or Spearfish suffer from back pressure due to the surrounding water, especially at great depth. This lowers power, increases fuel consumption, and puts extra stress on the engine. The reciprocating engine suffers less than the turbine, but both have limited development potential. If greater performance is required, and submarine development will probably enforce it, a closed cycle system is desirable. This might take the form of a stored chemical energy propulsion system (SCEPS), using lithium and sulphur hexafluoride as fuel. The heat generated by the combustion of these substances can be used to create steam to drive a turbine on a closed Rankine steam cycle. Alternatively, advanced battery and electric motor technologies promise similar potential power outputs. A lithium/thionyl chloride battery might yield almost twice the amount of energy per unit weight than a SCEPS system. Batteries, however, have less development potential and there are also problems to be solved with high currents interfering with electronics, and explosion dangers due to high electrical density. Future heavyweight torpedoes using advanced propulsion systems, and with improved hydrodynamic shape, might be able quietly to engage targets at virtually any operational submarine depth out to 50 km and at speeds of 80 knots. Wake-homing techniques have ASW as well as ASUW potential.[22]

Heavyweight torpedoes have the great advantage of a large warhead which can cope with the newer types of tougher, double-hulled submarines. It is a weakness of aircraft and surface ships that they usually have to rely on smaller 'lightweight' torpedoes for ASW purposes. These are usually autonomous weapons, being dropped over the target or fired in the correct direction from an on-board torpedo tube. Based on information from dipping sonar or sonobuoys, of from a ship's sonar, the torpedo then begins a programmed search pattern using sonar, first to detect and then to 'home on' to the enemy. Such a weapon needs to be able automatically to adopt the optimum sonar beam for the environment, searching out to the maximum range in minimum time. The weapon ought to be able to have sufficiently powerful electronics to cope with enemy tactics, evasion techniques, and countermeasures as programmed into its 'memory'. All this must be crammed into a size less than half that of a 'heavyweight' or 'long' torpedo.

The most widely used lightweight torpedo in the West is the 324 mm American Mark 46, later versions of which use the Otto propellent of their larger counterparts to drive them at up to 40 knots and up to 11,000 metres. The active/passive acoustic head has an acquisition range of about 450 metres.[23] Lightweight torpedoes need even more efficient power plants than their larger cousins, and they have pioneered the use of advanced closed cycle systems. The new American Mark 50 has a SCEPS power unit, giving speeds of up to 55 knots and ranges of 15,000 metres. The British are developing a SCEPS/Rankine cycle system known as CCTS (Closed-Cycle

Thermal System) for possible use in a future lightweight torpedo. Advanced battery technologies, with their high specific outputs, also have obvious application in lightweight torpedoes. The current British Stingray also already uses a magnesium silver chloride battery which gives similar performance to the Mark 46's Otto system, but more quietly and down to greater depths.[24]

Britain's Stingray and the American Mark 50 are good examples of the way a torpedo's guidance system can be improved by the application of more advanced computer software. These increasingly 'smart' weapons can better distinguish targets from clutter or countermeasures, select the optimum homing mode for the conditions, and be adjusted to cope with changes in the enemy's tactics or countermeasures.

France and Italy, who have produced rather crude lightweight homing torpedoes, are both now developing weapons comparable to Stingray in guidance and propulsion. Sweden has its own lightweight torpedo, the Type 42, a wire-guided passive homer specially designed for ASW and ASUW use in the Baltic from a wide range of platforms. The Swedes face sonar problems in these shallow waters, which have discouraged the use of the active mode in homing torpedoes. But the advent of quieter submarines is forcing a move in this direction, one aided by modern electronic and computer control.[25]

The Soviet Union utilizes electrically-powered, acoustically-homing, heavyweight torpedoes from its submarines and surface ships for ASW. Long-range wake-homers may also have a role here. It also deploys a somewhat larger lightweight torpedo than any other producer: a 406 mm weapon carried in aircraft, smaller surface ships, and submarines. Earlier models were passive only, but the current type is an active/passive device with wire guidance.[26]

To protect itself from torpedoes the submarine can either use decoys or make itself physically hard to destroy. Bubble generators, intended to deceive an enemy into thinking that a submarine has been destroyed, can act as underwater chaff, while torpedo-type underwater 'training targets' simulate submarines' acoustic characteristics. Specialized torpedo acoustic decoys known in America as ADCs (acoustic device countermeasures) are available to be ejected from the submarine when it is attacked, and hang in the water specifically programmed to jam the threat. Future anti-torpedo torpedoes also have relevance to submarines as well as surface ships. If a hit cannot be avoided a double hull, as used in recent Soviet submarines, may offer protection. Specialized penetrating warheads go some way to solving this problem for the attacker, although they provide an extra challenge to the guidance mechanism to deliver the weapon at the correct angle. The problem of lethality, coupled with increased speed and performance of the potential targets, casts a real shadow over the long-term viability of the smaller torpedo which, as Norman Friedman has pointed out, would have troublesome operational effects given the importance of these weapons in the armouries of the world navies. Certainly the challenge of developing all kinds of torpedo to defeat ever more formidable countermeasures is going to limit production to an even narrower range of states than previously.[27]

Homing ASW torpedoes can also be delivered by 'stand off' weapons of various kinds. The United States and its allies deploy the anti-submarine rocket (ASROC) missile system on surface ships, which uses a ballistic rocket to send a Mark 46 torpedo over several kilometres out to the computed position of the submarine. A new vertical-launch version will soon be deployed. Australia has developed a guided missile, Ikara, to deliver a similar payload over a rather longer distance (20 km). Both will be used for some time to come and similar systems are under development. The Soviets deploy a broadly similar, but larger, system as their primary ASW surface ship armament. The SS-N-14 has an estimated range in the region of 50 km. The Soviet Union also deploys a torpedo-carrying, underwater-launched ASW missile in submarines. The United States, which has so far only used this technology with nuclear warheads, plans to fit a Mark 50 homing torpedo on its 185 km (100 mile) range 'Sea Lance' ASW-SOW (Anti Submarine Warfare Stand-Off Weapon), which will be deployed aboard submarines. Given adequate communications this could be targeted on the basis of sensors and computerized data bases external to the launching platform, but doubts have been expressed over its ability to engage targets at maximum range without a nuclear warhead (which was delayed indefinitely in 1986).[28]

An alternative to taking the torpedo to the submarine is to get the submarine to come to the torpedo. The ubiquitous Mark 46 acts as the kill mechanism of the CAPTOR ASW mine, laid in deep water by aircraft, submarines, or surface ships. The mine listens for the distinctive sound signature of an enemy submarine and when it detects a target (at up to 1,000 metres) it ranges on it, using more accurate active sonar. When the target is within range the torpedo is released. The Soviets have rapid rising ASW mines of their own reportedly utilizing some form of underwater jet/rocket propulsion and either passive/active acoustic or UEP detection and homing. Honeywell and Plessey are developing the Crusader Advanced Sea Mine that is a deep-water rising weapon that can be laid beyond the continental shelf.[29] As an alternative to acoustic homing, ASW mines can be activated by the passing submarine's electrical field. All these mines are designed for use off known enemy submarine bases and in known transit 'corridors'. Submarine operators, such as the British, are having to emphasize special means of sweeping such weapons from relatively deep-water areas.

Mines have a long pedigree of ASW devices, as do depth charges, the first specifically ASW weapons to be developed. They were first of all dropped over the stern or fired by mortars over the side, but World War II saw the development of various 'ahead throwing' systems which allowed for a better interface of sensors and weapons, and a more controlled engagement. Both rockets and mortars were utilized in this manner, and still are in many of the world's navies.

Depth charges have several advantages still. They can be dropped to exert pressure on a submerged target rather than with the intention of sinking it. They thus have a graduated response capability that a more deadly homing torpedo lacks. Although they have little capability against, for example, a nuclear submarine, the threat of underwater explosion can be enough to cause a commander to move away, e.g. HMS *Conqueror* in 1982. Similarly they have the flexibility to damage a submarine that is too close to the surface to be vulnerable to lightweight homing torpedoes, as in the case of the Argentine submarine *Santa Fe* when attacked by British helicopters in the same conflict.[30]

For some time rockets or projectors for hydrostatically exploded depth charges have been replaced by lightweight torpedoes on most modern surface combatants built in the West, although most navies are retaining the capability to drop depth charges from helicopters or other aircraft. Older surface vessels, however, often continue to carry them using either racks or more sophisticated projectors, like the excellent British Mark X mortar, capable of delivering three accurately fused depth bombs in any direction from the ship out to 1,000 metres. The French still deploy an interesting 305 mm mortar in their older frigates, with a range of 3,000 metres, which can be used in a secondary shore bombardment role. It will take some time for such systems to disappear completely and naval commanders may miss them sufficiently to order their return. The weight of the charges is certainly a potential counter to even the new generation of tough submarines.

Still in production are various ASW rocket launcher systems, descendants of the British Hedgehog of World War II. These project a number of rockets, fused for contact, time, proximity (hydroacoustic), or depth. Those navies with shallow water ASW problems continue to be specially interested in these systems. The Norwegians carry the Terne II sextuple ASW rocket launcher in their destroyers and corvettes. This fires six time- or proximity-fused rounds out to a range of 900 metres against targets whose range and depth have been computed. The 48 kg warheads have a lethal distance of about 20 feet against a normal submarine. The Swedish firm of Bofors produces a widely exported tubed rocket launcher for 375 mm proximity/time/direct action rockets. A French sextuple variant was fitted to new, small coastal escorts of the French and Belgian navies built in the late 1970s and early 1980s. Japan continues to fit quadruple launchers of this type to its latest frigates, in addition to lightweight torpedo launchers.[31]

An especially interesting variant on this theme is the ELMA system developed by Sweden to inflict limited damage upon peacetime underwater intruders into her territorial waters. This consists of a number of multiple launchers for small, shaped charge grenades which are scattered in the chosen pattern and quantity over a range of about 250–350 metres. The shaped charge projectiles are designed to drive a narrow 30–50 mm hole into the hull of the delinquent submarine thus forcing it to the surface. This 'imaginative' limited response to the problem of the covert underwater intruder is in service on Sweden's fast-attack craft.[32]

The Soviets, with their tradition of military rocketry dating back to World War II, remain the widest users of ASW rocket launchers. The projectiles are fired on to a contact's position, using either depth or influence fusing. The standard 6,000 metre range version is carried in all Soviet surface ships from Petya class corvettes to Kirov class 'battlecruisers', and smaller vessels carry smaller variants. Such devices will give the Soviet Navy considerable combat flexibility in shallow water ASW for some time to come.[33] ASW is an activity fraught with so many uncertainties that all kinds of weapons will have a part to play in countering the invisible and increasingly quiet menace of the modern submarine. Indeed, it seems likely that the increasing quietness of the new targets, both nuclear and conventional, will reduce the effectiveness of the passive sonars upon which most ASW techniques have relied in recent years.[34] Large, low-frequency, long-range, passive sonar arrays have long been mounted on the ocean floor in the American SOSUS (Sound Surveillance System). This is being supplemented both by denser, shorter-range static arrays covering certain areas (the so called Fixed Distributed System) and mobile strategic surveillance arrays towed behind unarmed surface ships known as T-AGOS. Combined with modern electronic communications analysis and dissemination techniques these systems have done much to deny noisy submarines their cover. By the 1970s this had already reached the extent that it was being publicized that Western naval commanders have constant access to the rough location of those potentially hostile submarines that were at sea. This has provided the technical rationale to the operational doctrines adopted and developed over the last thirty years by NATO. ASW has tended to be viewed as a kind of air warfare with targets detected at long range and platforms directed accordingly to deal with them. These latter assets also rely on long-range passive detection for much of their capability, in particular the tactical towed arrays streamed by both submarines and surface ships with ranges of many tens of kilometres.[35]

As the number of submarines with quiet machinery increase there will be less low-frequency machinery sound to penetrate long distances in the ocean carrying its wealth of information to the passive signal processing equipment of the whole ASW system. Informed analysts now argue that ASW in the future may well have to re-emphasize much shorter-range passive detection opportunities. Ranges of 50 km or so (the first convergence zone) may become the maximum realistically attainable in all except the most favourable circumstances.[36] Often ranges will be much less. Nevertheless, the tactical towed array for the time being remains a crucial long-range detection asset and it has allowed the surface ship to reassert a new importance in ASW. Surface ships can carry longer, more capable arrays and, unlike submarines, are freer to share their information with other forces. Aircraft can be quickly 'cued' in to prosecute possible contacts using shorter-ranged sonobuoys or dipping sonars. There is no doubt that much work will continue with signal processing, e.g. in hitherto little used parts of the noise spectrum, to retain the long-range capability of the passive towed array.

The performance of more traditional active/passive hull-mounted and variable depth transducers can be further improved by means of still greater improvement in digital electronic signal and data processing. Convergence zone active detection and engagement should, therefore, still be possible for these systems against even the quietest submarine in good water conditions, at least. The data processing challenge, however, is enormous. A modern active sonar gives around 4,000 pieces of data per second, a rate that generates *false* alarms every 2–5 seconds or so. In a passive sonar the rate is over 60,000 items per second, giving a false alarm every 15 milliseconds. This makes automatic 'target extraction' of particular importance. One possibly significant development is the use of combination active transducer/passive array systems. These allow long-range sonars to be fitted to small vessels unable to carry a large transducer array themselves.[37]

The combination of active 'transmitters' and separate passive array 'receiver' technology will be the key to new, low-frequency active sonar systems that will be developed as the wholly passive sonar window slowly closes. These will take novel 'bistatic' or 'multistatic' forms with static, towed, or hull-mounted sound sources sending out omnidirectional signals to be reflected by the target and received on long-range passive arrays either static or towed behind other units. Such systems have been proposed, in both Europe and the United States. Very long-range, strategic, very low-frequency systems as well as shorter-range tactical systems are conceivable, like today's wholly passive arrays. As C.C.F. Adcock put it to the 1988 Undersea Defence Technology Symposium:

> In general, remote receivers can yield additional areas of possible detection of submarines outside the area that would be covered by a monostatic receiver located close to the sound source. An additional advantage is that, although alerted to the sonar transmission, submarine commanding officers have no means of discovering the location of the acoustic receivers. They, therefore, lack the necessary information to be able to manoeuvre to keep away from receivers, and present low target strength or minimal doppler. 'Beaconing' (i.e. revealing the transmitter's position), might also be less dangerous if the signals were made 'broad band' and thus hard to distinguish from normal background noise. Beaconing, however, is not the only problem; care would also have to be taken not to give away the position of one's own units which would also reflect the radiated sound. The tactics of long range, low frequency active sonar will pose a fascinating set of conundra for the tactician of the future.[38]

While surface ships and submarines equipped with active sonars are obviously vulnerable to beaconing and reflection, an aircraft is less so. Longer-range, low-frequency, active dipping sonars carried in helicopters are, therefore, likely to be one key to maintaining effectiveness in ASW; in suitable water conditions convergence zone ranges may be possible, ten times the 5,000 metre range of current medium frequency systems. One such system under development is the Anglo-American HELRAS (Helicopter Long-Range Sonar).[39]

Closing the passive sonar window is bound to limit the utility of the air-dropped passive sonobuoy. Since 1960 low-frequency analysis and recording (LOFAR) buoys, using the Jezebel analysis capability developed in the 1950s for SOSUS, have allowed detection at ranges of up to 100 nautical miles. Over the next fifteen years DIFAR (directional LOFAR) buoys were developed and the United States has vertical line array DIFAR (or VLAD) buoys for deep surveillance. Extending the capabilities of these systems

Weapon systems

will be necessary, with improved signal processing capable of monitoring larger numbers of buoys of more varieties. Active buoys, always important for target acquisition, will however inevitably acquire a new relevance with new designs both to utilize better batteries, and to make better use of those power sources. Rotary winged aircraft will probably also be fitted with control systems designed to make integrated use of sonobuoys and dipping transducers. Larger passive sonar arrays that can be dropped into the water and then recovered may be practical as 'Osprey' type aircraft gain more general acceptance (see Chapter 6.)[40]

Submarine launched weapons with fire control solutions based on what will still often be the best sonar information, co-ordinated by increasingly sophisticated and integrated command and control systems and carrying the most effective warheads, are probably going to remain *individually* the most effective ASW weapons. Nevertheless, the developments just described will emphasize the multidimensional character and complexity of anti-submarine warfare.[41]

Anti-air warfare (AAW)

Ever since the invention of the aircraft, the threat it poses has been a central concern for naval officers anxious about the survival of both their forces and the objects of their protection. The more recent advent of the missile has made the problem even more serious as this has diversified the threat (submarines can now pose 'air' problems), and created the need to destroy the threat physically rather than just put if off its aim. Much of the earlier effectiveness of AA gunfire was rather that of the modern chaff dispenser, putting the enemy's 'guidance system' off its aim. There was greater scope for effectiveness as a man's determination is less predictable than that of an automatic system. Now and in the future, however, AA systems must inflict catastrophic material damage if they are to be effective.

This is not to say that traditional systems like guns, even those without very sophisticated fire control, cannot be effective against a range of air threats. A tendency to define the air threat too narrowly in technical terms, to emphasize long-range aircraft and the missile rather than small aircraft armed with bombs, led the Royal Navy for one to remove light AA armaments from ships in the 1950s and 1960s and rely on dual-purpose medium guns and missiles. That this was insufficient to cope with the air threat disposed by a relatively weak power like Argentina in 1982 demonstrated the shortcomings of this policy. Light guns in the 20–40 mm bracket have made a dramatic come-back and will probably remain. They can at least make a close aerial attacker's task much more difficult, and more advanced fire control and sensors ought to enhance effectiveness still further.

At the upper end of the light gun spectrum come the sophisticated anti-missile gun systems combining advanced autonomous automatic radar tracking with ultra-high rates of fire. These are also going to become standard fitments to the ships of the more sophisticated navies. The Vulcan Phalanx CIWS (Close-In Weapons System) is widely deployed in the US Navy, and alternative European systems are available in the form of Goalkeeper and Seaguard. Such weapons fire ultra-high-velocity kinetic energy rounds designed physically to destroy the incoming weapon. These guns cover the gap in the minimum ranges of many AA missile systems, and are effective against current missile threats. They are, however, relatively expensive and the extent of their deployment is thus limited. Their effectiveness against future faster missiles is also in some doubt. Larger artillery in the 114–127 mm bracket is also likely to have a continued rule, albeit a relatively subordinate one for the AAW task. The US Navy has experimented with infra-red guided shells for anti-aircraft work and stayed with the more rapid firing 5 inch (127 mm) gun, largely due to its AA potential. The 4.5 inch (114 mm) weapons in British ships did score some isolated successes in the Falklands and added to the generally hostile environment faced by Argentine pilots.[42]

Despite the continued relevance of guns, rocket propelled missiles seem to be the key instruments of short-range air defence. This revolution has been longer in coming in reality than it might have seemed in theory, given the extent of missile deployment in ships. It must be stated, however, that most of the first generation systems were so limited in capability that their real combat effectiveness was highly dubious. At best the longer range systems could destroy individual high-flying aircraft, e.g. long-range bombers. The shorter-range systems stood some chance of dealing with a closer-range attack, although one of the most widely deployed, Seacat, has stood persistently low in the esteem of its operators, at least in its earlier versions. A simple radio command weapon, it depends very much on the skill of its human operator. Seacat itself has been improved, both in sensor and control technology, and the missiles are now much more reliable and effective, and are likely to become even more so. Nevertheless, it is significant that the most widely-deployed first generation point defence missile systems tend to rely on more automatic semi-active homing radar techniques, usually using air-to-air missile technology.

Anti-air warfare operates in an easier medium than ASW, but the speed of reaction required makes it complex enough. We may be approaching a situation where very few navies will be capable of deploying systems able to cope with air threats of the highest level in both quality and quantity. The only such system available currently is the American Aegis, which combines phased array radars with the most advanced electronic data handling techniques. The system can cope with many targets and, in its initial form, launch about thirty-four missiles per minute. These are command and auto-pilot guided into the vicinity of the enemy, on whom an illuminator is switched to provide the necessary reflected waves for final homing. More effective homing trajectories allow a greatly extended range (by 60 per cent) for any given missile. Aegis is currently deployed with the Standard SM-2MR missile, giving it an impressive maximum range of about 150 km (90 miles). It can track targets out to 370 km (200 miles). The system is highly resistant to electronic jamming and offers formidable defensive potential – at equally formidable cost. It is currently only deployed by the US Navy in its Yorktown class cruisers, of which twenty-seven are planned in total. Aegis is also to be fitted in modified form to the new US class of Arleigh Burke destroyers now under construction, and Japan plans to put it into its four new 6,000 ton destroyers.[43]

The European navies seem to have been put off not so much by sheer cost as by the industrial implications of not being able to interface European hardware with the expensive software that is the core of the system. However, the multinational anti-air warfare system currently being proposed for the NATO frigate of the 1990s may be based around Aegis, which might allow European navies to get their hands on the vital software cheaply. France, the major opponent of an Aegis software buy, has reportedly been assured of the compatibility of Aegis with its new Aster 90 vertically launched surface-to-air missile. In its initial Aster 15 form this is a relatively short-range system (10 km against missiles, 17 km against aircraft). France hopes to develop a longer-ranged Aster 30 as an area defence weapon, but a general trend in the European navies seems to be towards pushing out the range of point defence missiles rather than emphasizing expensive new area defence systems. The new vertically launched Sea Wolf is another example of this kind of development. The use of inertial navigation systems in these new missiles results in more efficient trajectories, longer range, and diminished demand on the ship's fire control system. Vertical launch allows very high rates of fire, one missile per second. Such projectiles could be combined with cheaper radar hardware: one British firm, Plessey, is claiming a breakthrough in gallium arsenide micro-circuitry to provide significantly smaller and less expensive 'multi-function electronically scanned array radars' (MESAR). The whole combination might well amount to an affordable combination of hardware and software that will allow European ships to operate in high threat environments. Co-operative procurement is, however, the key. It is unlikely that any single national European navy could afford its own system of this level of capability.[44]

As one might expect, the Soviets seem to have some difficulty in coping with the computer technology associated with this level of capability. The only system in any sense comparable is the SA-N-6 of the Slava and Kirov classes. In these two classes the relatively large missile is controlled by a large Top Dome phased array radar that is steered in azimuth, unlike the fixed multiple radar faces of the Americans. According to one American critic this 'suggests a limited ability to handle threats from different relative positions simultaneously'.[45] While the Kirov class carries two such radars, the Slava class carries only one, which must limit further its all-round coverage. The only Soviet ship with electronically scanned phased arrays similar to those used in Aegis is the fourth aircraft-carrying cruiser *Baku*. Oddly, however, this vessel does not seem to carry a compatible area defence missile system. The phased array radar's role seems to be fighter direction and air warning.[46]

Moving down the scale, the world's navies are well equipped with missiles which use semi-active homing to attack an incoming aircraft: the missile homes on the reflected energy of a ship's radar. The longer-range variants cover a spectrum from about 30 km to over 100 km. Sea Dart, the British missile designed to give long-range AAW protection, gave excellent service in the Falklands War against well-defined aircraft targets detected in good time against a clear background. It destroyed six Argentine aircraft and (by mistake) one British. It was predictably less effective, however, against low-flying targets coming out from the coast, as HMS *Coventry* found to her cost.[47] Another general problem with weapons of this type is their ability to engage only the number of targets for which they have illumination radars. The capacity of the system can be improved by giving a mid-course guidance capability to the missile, as the Americans are doing with the long-range variants of the SM-2 missiles deployed on their older missile cruisers. This, however, once again implies expensive modifications to the ship's missile control system.[48]

Whatever their capability of dealing with multiple targets, such systems force enemy aircraft to attack from low altitudes, so lowering range and weapons carrying capacity. Low-altitude targets that get through the longer-range missile envelope are dealt with by shorter-range systems. These may be advanced command-guided

Weapon systems

systems using radar and or TV trackers, like the British Sea Wolf. The latter proved successful in the Falklands (four aircraft and a contribution to a fifth), although prone to confusion at first due to narrow programming of its software. It will provide the vital self-defence capability of British warships into the next century. With the advantage of combat testing it is now probably as effective a close-range defence system as any in the world. The early variant, however, was only intended for short-range use (up to 5–6 km) and could only be used in support of other units at some danger to the platform due to the close manoeuvring involved, and at some risk of tactical confusion.[49] Moreover, its fit to all ships was constrained by cost and relatively high topweight demands. The latter problem, at least, has been dealt with by the creation of a lighter launcher and also the new variant for vertical launch, which also, by giving it extra boost, significantly increases its range and flexibility and converts it into something that can protect something else as well as just the platform carrying it.[50]

Sea Wolf was designed to deal with the Soviet missile threat. Less capable point defence systems provide adequate defence against low-flying aircraft. Missile technology designed for land-based surface-to-air or air-to-air use is often used with various forms of guidance: semi-active radars, infra-red, command-to-line-of-sight. They can give vessels down to fast-attack craft a useful self-defence capability. They are produced for export and domestic use by most of the arms manufacturing nations, including Israel, and should increase in effectiveness incrementally over the years.

It is possible that, in the future, missiles might acquire longer ranges and greater effectiveness still by the adoption of some kind of external control system whereby a surface vessel is only the launch platform and the engagement is controlled by an airborne or even satellite platform. This has been discussed in the United States as, in American football parlance, a 'forward pass' or SLAT (surface-launched air-targeted) missile. Developing such a system would be very expensive and problematical, especially in terms of adequate data links. The link between a missile and its own launching platform is much more secure. Nevertheless, work continues on such concepts, and the increasing adoption of vertical launch systems with the potential for loading large, long-range missiles gives the idea some credibility.[51]

The last way of dealing with air threats at maximum range is to use aircraft of one's own, which raises bureaucratic and organizational as well as command and control difficulties (see Chapter 6). The use of aircraft as missile launch platforms makes air interception especially lucrative. Carrier-based fighters, combined with air-to-air missile technologies, are still vital weapons if one can afford them. One central lesson of the Falklands War, however, was the effectiveness of relatively crude and simple fixed-wing platforms (Sea Harriers), when combined with advanced infra-red homing missiles (AIM-9L Sidewinders). Possession of the superior missile deprived Argentina of any hope of dealing with the main defence of the British Task Force from air attack.[52] The development of the new types of active radar guided missiles, like AMRAAM (advanced medium-range air-to-air missile) will enhance the capabilities of such relatively unsophisticated aircraft as the Sea Harrier still further.

The main constraints on the operations of the Harriers in the Falklands were their minimal radar capability and, more importantly, the lack of airborne early warning radar coverage. Land-based aircraft can provide the latter within range, but adequate results can be obtained by radars mounted on helicopters. It would seem prudent not to deploy short take-off vertical landing (STOVL) fighters without such AEW helicopter support. The combination is obviously less capable than a fleet air defence envelope of US proportions, but only the US Navy is in that kind of league – a circumstance that will not change for many years.

The level of air defence capability provided by the combination of Phoenix missiles, F-14 Tomcat fighters and E-2 Hawkeye AEW aircraft provides protection that stretches well over 500 km around the centre of any carrier battle group.[53] Each aircraft can launch up to six Mach 4 Phoenix missiles, each of which can be directed on to a separate target at ranges of 120 km using a combination of semi-active and active radar homing using the aircraft's own radar as illuminator. Phoenix was primarily designed to deal with Soviet missile-carrying bombers. Lesser missiles, both radar and infra-red guided, are more appropriate for close-range encounters with fighter type aircraft, and can be carried by the F-14 or its less capable F-18 stablemate. The ability to fight for air superiority is formidable, and US carrier battle groups can think in terms of controlling the air environment in which they find themselves to a considerable degree, even to the extent of taking on the Soviet Naval Air Force close to its bases.[54]

It is in AAW that the naval warfare of the future may get into the realms of the exotic. Ultra-high-velocity railguns might well have an application given their potential hitting power, and speed of response and engagement. The availability of electrical power from the ship's machinery should make their mounting in naval platforms easier than in the tanks for which they have been proposed. As for directed energy weapons, relatively low-powered electro-optical lasers of sufficient power to blind and/or damage sensors, including pilots' eyes, are currently practical weapons and are already deployed at sea.[55] As work is carried out by both superpowers on more powerful directed energy weapons as part of anti-ballistic missile (ABM) research and development, these very advanced technologies might well become available for the defence of warships. In the short term, however, the American Strategic Defense Initiative may have slowed down the development of exotic naval weapons. The US Navy's Chair Heritage programme to defend carriers with particle beams was absorbed into the SDI and developed a new ABM emphasis. All directed-energy weapons have one enormous advantage. By working at, or close to, the speed of light they require few sophisticated fire control calculations, just accurate pointing.

Operations against the shore

Admiral Gorshkov has pointed out that perhaps the main revolution in modern naval affairs has been the greatly enhanced utility of navies for operations against the shore.[56] Bombardment is a long-standing role, either for its own sake or in support of armies. Guns, aircraft-launched weapons, and missiles will all continue to have a part to play in this form of warfare.

Coastal bombardment, even more than ASUW, will be the major role of the medium/heavy calibre artillery carried in ships. Vietnam confirmed the US Navy in its continued requirement for such capabilities, as did the Falklands War for the British. Guns can deliver fire more accurately than aircraft, in greater quantity and with less risk to personnel. This is particularly true in limited war and gunboat diplomacy situations as it is more likely that the object of the exercise of force will be better protected against aircraft than seaborne long-range artillery.

The US Marine Corps found especially welcome the reactivation of the Iowa class battleships, as their 16 inch guns make formidable shore bombardment weapons. The accuracies obtained with the old propellent available initially proved to be less than expected but these problems have now been solved by remixing. Even the old armour-piercing rounds might have a role against some targets, such as airfield runways, and cluster munitions offer considerable potential in both the anti-personnel and anti-armour role.[57] Nevertheless, it is hard to see new guns of this calibre being deployed given the adequacy of modern medium artillery in the 5–8 inch bracket to cover realistic bombardment tasks. US ships still mount 5 inch (127 mm) guns, and the Soviets also mount nothing larger than 130 mm (5.1 inch) guns in their new ships, even their new battlecruisers. Calibres may well creep up, however: the British are developing a

The changing shape of naval war

155 mm weapon, based on army field howitzer technology in order to provide greater explosive firepower and, more importantly, an improved facility for indirect fire in mountainous (fjord-like) terrain. It would also not be surprising if the Americans eventually returned to something like their 8 inch Major Calibre Light Weight Gun project, abandoned in 1978 (in part to help the battleship case), together, perhaps, with a specialist shore bombardment 'fire support ship' to carry it.[58]

For shorter range, multiple rocket launchers offer many advantages in their ability to cover broad areas with fire. They are already in use to some extent and there would seem to be few problems in applying the new guidance and warhead technologies being considered in connection with NATO's new multiple-launched rocket systems to shipboard variants. This could allow even more formidable fire support to be delivered in support of amphibious forces.

Missiles can be used to carry out deeper conventional bombardment of land targets. The American Tomahawk cruise missile has two conventional variants TLAM/C and TLAM/D for this task. The former is coming into service, the latter will be available by 1990. These missiles have a range of around 1,000 km and navigate themselves over land using terrain contour matching (TERCOM) guidance, updating a basically inertial system by comparing a digital picture of the ground over which the missile is flying with a 'map' in the missile's computer memory. On closing the target a digital scene matching area correlator (DSMAC) system compares an optically-scanned picture of the target area with another digital picture in the computer memory. This increases accuracy to about 30 feet to give TLAM/C's single 972 lb blast fragmentation warhead a chance of inflicting serious damage on a precise target. TLAM/D's submunition warhead will add to the Tomahawk's conventional effectiveness, and experiments have been carried out with one missile making a number of bomblet attacks on separate targets. Nevertheless, conventional Tomahawks will still probably have to be fired in large numbers to be sure of achieving significant military effect. This means that platforms with large magazine spaces, i.e. surface ships, are necessary for it. Submarines simply cannot carry enough missiles. It is hard to imagine suitable targets for individual missiles, especially the current TLAM/C, which is a relatively costly way of delivering a very limited amount of ordinance. Pictures have, however, been published of successful attacks on parked aircraft and bunkers, and it might be useful in making exemplary 'anti-terrorist' punitive strikes, especially in circumstances when loss of pilots would be embarrassing. Used in adequate numbers the conventional Tomahawk's major role is defence suppression in support of more

Weapon systems

conventional air strikes. Radar and missile sites, and control centres, would be lucrative targets and even airfields might be thrown into confusion, if not knocked out physically, by a series of massed TLAM/D strikes, especially if missiles were fitted with specialized submunitions, perhaps with time fuses.[59] Mike MccGwire has suggested that the new Soviet SS-N-24 large sea-launched cruise missile currently being tested on a converted Yankee class SSBN (Nuclear Powered Ballistic Missile Submarine) may be a conventional long-range bombardment system for use against command centres and nuclear assets in Europe in the early days of a conventional war.[60]

Sea-based aircraft provide the final and most important means of carrying out conventional 'naval bombardment'. The capabilities of the US Navy to carry out attack missions with a wide range of conventional munitions ('smart' and unguided) is as good as any land-based air force's, although limited numbers of all-weather strike aircraft can be a constraint in certain circumstances, e.g. in the attacks on Libya in 1986, when attacking the full target would have required two sequential waves, both politically and operationally undesirable. Stealth attack aircraft are under development for the US Navy. All navies with organic air components have some ability to mount bombing raids, although no other navy is anywhere near to US capabilities. In the Falklands War, for example, the limited bomb-carrying capacity of the British Sea Harriers forced the use of land-based, long-range bombers, at considerable expense and to dubious effect. STOVL aircraft, however, can still be extremely useful, especially for the close support of ground operations.

Command and control

The twentieth-century revolution in command and control – the conversion of individual weapons into integrated weapons systems – is perhaps the least appreciated aspect of modern naval operations. Ships and weapons are too often compared by their outward appearance, and their performance parameters narrowly and simplistically defined. The less tangible elements may, however, be the most crucial. As early as World War I the performance of the British Fleet's gunnery was far more affected by the defective performance of its mechanical analogue fire-control computers and the interface of those computers with accurate but low-data-rate sensors, i.e. rangefinders, than it was by marginal advantages in numbers of guns, their calibre or even the quality of their projectiles.[61] In World War II the advent of truly three-dimensional naval warfare and electronic sensors that increasingly replaced the eye as the main means of identifying the enemy's position enforced a fundamental development in the way in which naval warfare was managed. Spaces in the ship had to be set aside for what in the Royal Navy was called the Action Information Organization (AIO), where the data from the ship's electronic sensors were integrated into a coherent picture. British practice was to call the surface warfare room the Operations Room. American warships had Combat Information Centers (CIC).[62] All this was space intensive and it put strains on the internal volumes of ships. Sometimes, this meant less obvious armament but what was left was much more effective. The 'heavy' armament of the ships of the past was a function of inaccuracy and inefficiency rather than real fighting power.

The data handling and collation initially had to be carried out by human means, but the advent of the electronic computer allowed mechanization and automation. The Americans began to computerize their CICs with Naval Tactical Data System (NTDS) in the late 1950s. This has been progressively upgraded since. In the 1960s the British followed suit and the carrier *Eagle* was fitted with ADA (Action Data Automation System). This 'co-ordinated air defence information from the ship's own radars, from other ships and aircraft by data link, and from visual sightings. This information was either stored in the computer's memory, or displayed. The computer was capable of evaluating threats, calculating interceptions and passing out information required by fighters, guns and missiles. It was able to perform tasks that normally took some sixty men.'[63] This kind of capability spread to smaller ships the following year with the ADAWS (Action Data Automation Weapons System) of the County class large destroyers. ADAWS was provided to various subsequent classes especially, but not exclusively, to those with an air defence role. It remains in service in various versions: two in HMS *Bristol*; four in the Type 42 destroyers; five in the Ikara Leanders; six in the first two ASW carriers; and ten in *Ark Royal*. A simpler system, known as CAAIS (Computer-Assisted Action Information System), was developed for frigates, and this has been replaced by CACS (Computer-Assisted Command System) introduced into the later Type 22s. CACS has not had a happy history. The first variant had software problems and CACS 4 intended for the Type 23s has had to be re-tendered, which will delay its entry into service until after its platform has been completed. The first Type 23s it seems will thus go to sea without their electronic 'brains' and hence much of their ability to fight.[64] History sometimes repeats itself.

The function of these systems has been described as follows:

> They aim to gather and collate information and present it intelligibly for decision. . . . They enable targets to be

Weapon systems

designated, and the two more comprehensive systems, ADAWS and CACS, directly control certain weapons, taking over the sort of function which otherwise is carried out by separate fire control systems.[65]

The captain of a modern warship now fights it not from the bridge, but from a centralized operations room in the bowels of the ship where officers and men pore over cathode ray tube displays making keyboard injections to input data or decisions. This revolution has made necessary structural changes to the naval profession, e.g. in the United Kingdom, where a new Principal Warfare Officer is now the master tactician responsible under the captain for the fighting of the ship.

Earlier AIO systems tend to be over-centralized and thus subject to battle damage. The revolution in microelectronics has, however, helped produce improved systems with spare main computers and operations decentralized to miniprocessors and microprocessors in the consoles. Capacity is increased, which speeds up processing and allows more inputs like passive sonar or electronic intercept information to be integrated with all the other tactical data. The use of colour screens and of light pens to interact with the screen makes systems easier for sailors to operate, and reduces the training load. The trend is to increase decentralization still further with smaller – and cheaper – computers in 'autonomous intelligent consoles' spread over the ship linked by protected data highways. Each part of this system could operate independently if required. Such modular systems also lend themselves to installation in small surface combatants and MCM vessels.[66]

Computerized AIO systems are now being designed for submarines whose sensors now produce so much data that automation of its handling is a necessity. The new British Submarine Command System (SMCS) now under development is designed to

> relieve the human link in the command of many of the operations currently necessary to provide the tactical picture. The Command is thus freed of the more routine and tedious operations which can be carried out by the computer, enabling the Command to concentrate more on the decision-making process which, in underwater warfare at least, is a much more complex and vital problem than in the surface or air environment.[67]

That very complexity, however, mandates a powerful and complex system, especially if a high degree of automation is required. The Americans have been facing serious problems with their SubACS (Submarine Advanced Combat System). The problem has been to integrate sonar information both in terms of automating passive detection and putting together the outputs of different sets operating on different frequencies over different acoustic paths, all in a system that has to operate in a restricted space.

The manifold limits to the human capacity to respond adequately on a consistent basis, the sheer data problems created by effective modern sensors, and the inherent complexities of modern three-dimensional naval warfare, mandate ever greater mechanization of tactical and command functions. 'Intelligent', knowledge-based systems displaying options to commanders will allow them to concentrate wholly 'on decision making rather than procedure management'.[68] Even certain decisions may be increasingly taken out of their hands. Already a degree of automatism is built into weapons control to cope with the rapidity with which threats can be presented. The human role will be increasingly to veto rather than to take the initiative. A major problem here will be that the ship may well be in a situation where the degree of discrimination required is beyond what can be fed into a computer, however advanced. Only when unlimited hostilities have begun can the man be left out of the command loop. Even then there might be inflexibilities in the command system's programming that an enemy can exploit. The human commander with his 'human faculties, such as complex forms of recognition, scepticism, realising the significance of apparently unrelated occurrences and exercising initiative and judgement' will be required at sea for a long time to come.[69]

Nuclear weapons

So far this analysis of technical trends in naval warfare has ignored the question of nuclear weapons. This is deliberate, as the use of nuclear weapons as an element in naval warfare lacks a certain plausibility. Beginning in the 1950s, the United States, Britain, and the Soviet Union all developed a wide range of nuclear weapons for use at sea: nuclear depth charges and nuclear-armed torpedoes to help solve the problem of the high speed submarine and make up for sensor inadequacies; air-dropped nuclear bombs and warheads for cruise and ballistic missiles to deal with large enemy surface combatants such as cruisers raiding NATO supply lines, or aircraft carriers threatening the Soviet Union itself.[70] The USSR even developed plans to call down long-range, land-based nuclear missile 'fire' on NATO naval concentrations, the most useful Soviet intercontinental ballistic missile of the 1960s, the SS-11 being originally designed for this purpose by the Chelomei naval missile design bureau.[71]

Weapon systems

Many of these weapons are still available (although the United States has unilaterally announced the withdrawal of its nuclear ASW and AA missiles) but their use raises a number of serious problems.[72] It is hard to see their relevance in other than a general nuclear war, in which naval operations would probably cease to be very important (see Chapter 9). Second, the weapons themselves would have most undesirable side effects that would impact on the user as much as the enemy. Nuclear depth charges would probably cause so much reverberation that they would seriously degrade sonar performance for some time after an explosion. Nuclear air defence would create considerable problems of electromagnetic pulse and ionization around a task force just at a time when clear sensor information was at a premium. These drawbacks are combined with the problem of political acceptability that nuclear weapons face when surface ships tend to be turned away from ports. The continued deployment of nuclear weapons in surface ships, if not carriers or submarines, is thus highly questionable. The conventional munitions of the future will be even more capable of carrying out the ASW and anti-shipping tasks than they are today, without resort to nuclear weapons. Indeed, to argue that nuclear weapons are necessary for certain tasks is to say that those tasks are impossible, as nuclear release will in all likelihood not be granted purely as a results of the dynamics of the war at sea. As long as one side in the superpower naval balance deploys nuclear weapons it would be imprudent for the other to get rid of all means of retaliation at sea, but this should not require a stockpile of any great size or the wide deployment of nuclear-armed systems that we have today.

Only two kinds of deployment of nuclear weapons at sea makes much sense. One is operations against the shore as part of one's overall deterrent posture. US carrier aircraft can deliver a range of nuclear bombs accurately over long ranges. TLAM/N variants of the Tomahawk cruise missile can convert any American nuclear attack submarine, or even a destroyer, into a potent strategic launch pad. TLAM/N is able to deliver a 200 kiloton warhead with an accuracy of 30 metres or so over a range of 2,500 km. The Soviets are developing a similar system, SS-N-21, as well as the larger, possibly dual capable SS-N-24 submarine launched cruise missile. Whether, given the problems of arms control verification, such systems are desirable strategic strike systems is another matter. They *could* have a special role as part of NATO's theatre nuclear forces in the aftermath of the mutual withdrawal of land-based intermediate nuclear forces (INF), but sea deployment makes them symbolically much less tied to Europe's security than if they were physically in the theatre themselves. Even if put under the command of NATO's

Supreme Allied Commander Europe (SACEUR), they would probably be seen as strategic and therefore as incredible of early first use as those submarine launched ballistic missiles already allocated to SACEUR. Moreover, the US Navy has always argued against any limitation on SLCM (sea-launched cruise missile) carriers' movements being involved in specific targeting plans, especially the overall strategic SIOP (Strategic Integrated Operational Plan). TLAM/N is the US Navy's own strategic reserve which it will remain most reluctant to share. It would be most ironic if converting their surface combatants into nuclear launch pads came back to haunt the US Navy, putting unacceptable constraints on their freedom of action. Navy attitudes to TLAM/N might well then change.[73]

The other naval nuclear role with some logical stature is the French one of use in *pré-stratégique* strikes to demonstrate resolve to initiate the use of national strategic forces. Large Soviet naval targets might be more suitable candidates for such desperate measures than land targets, and there is thus some rationale for plans to procure in 1988 twenty-four ASMP 300 kiloton, 250 km range, inertially guided, ramjet powered missiles for the Aeronavale's carrier-borne Super Etendards. The latter are already configured to carry the 20–25 kiloton AN–52 free-fall bomb. The light carrier *Foch* came into service as an ASMP capable ship in the summer of 1988, and the new nuclear-powered carrier will also field the weapons.[74] One wonders whether the nuclear bomb capability announced for Britain's Sea Harriers reflects similar thinking; it is hard to think of other reasons for so configuring these short-range aircraft in this role.[75]

Most important in general strategic terms, and the main reason that at least one superpower (the Soviet Union) needs to be able to use the sea, is the deployment of submarine-launched ballistic missiles (SLBMs). Solid, or in the Soviets' case, storable liquid fuelled, long-range ballistic missiles carried in submarines are acquiring ranges almost comparable to land based ICBMs (over 7,000 km or more). They can, therefore, be fired by one superpower at the other from waters close to home. Their throw weights are also becoming sufficient to give them a generous loading of multiple independently targetable re-entry vehicles with a large footprint, i.e. an area in which an individual missile can attack separate targets. All five current nuclear powers deploy SLBMs and, if not capable of any other modern long-range nuclear delivery mode, deploy them alone.

Traditionally, SLBMs have not had a sufficiently high combination of accuracy and yield to attack hardened targets, such as missile silos or launch control centres. They have, therefore, been directed

Weapon systems

against soft military targets and urban industrial centres, sometimes whole areas. The apparent British plan to direct all sixteen of their deployed Polaris missiles, each with 2–3 warheads, at one area, Moscow, is a somewhat extreme variant of this thinking.[76] Now, however, the combination of yield and accuracy of SLBMs is approaching the point where they will be usable to attack hardened targets. The US Trident D-5, planned for the 1990s for deployment by the US and Royal Navies, will be able to place its eight 475 kiloton MIRVs accurately enough to take out the hardest target. The CEP (circular error profitability – the radius of a circle in which 50 per cent of the warheads are expected to fall) is about 250 metres, similar to current ICBMs.[77]

This enhanced accuracy is not just a result of missile improvements, including updating the inertial guidance system using star sights (the latter technique is also used by the USSR), or NAVSTAR (navigation system using tuning and ranging) satellites, but also by providing the launching submarine with means of updating its own inertial navigation system. Trident submarines will be able to scan the ocean floor, rather as a cruise missile scans the ground over which it is flying, and, to the same purpose, to update the on-board inertial navigation system to achieve pinpoint accuracy. Naval forces will thus soon be physically capable of carrying out the ultimate operation of modern war, the full range of strategic nuclear attacks against the homeland of the other superpower.

Chapter 6

Platforms

By naval 'platforms' we mean the vehicles by which military power is exercised at and from the sea. These have for a long time involved much more than ships. Aircraft and submarines are now as important parts of the maritime environment as surface ships, sometimes more so. Space platforms are already making an impact at sea, and may do so to an increasing extent in future. But before we move to assess the importance of each of these means of exerting force at sea, we must look at the projected evolution of the main mode of sea use that these platforms will be protecting – namely the *ships*, designed to carry materials of greater human and economic wealth than the instruments of destruction. Shipping development sets the context for naval development.

There can, as argued elsewhere in this book, be little doubt that the sea will remain the world's most important long-distance cargo transport medium and that sea transport will continue to be man's primary purpose in 'going down to the sea in ships'. It seems unlikely that for the foreseeable future those ships will be anything else but the familiar 'displacement' surface vessels, floating at the sea/air interface. The now familiar mix of tankers, bulk carriers, container ships, and general cargo ships will continue to carry the world's seaborne trade. Economic factors more than technical ones will probably prevent any major revolution in design or characteristics that will fundamentally affect the current situation. The displacement ship is simply the most efficient way of carrying material, and only in certain specific cases where quite literally 'time is money' (such as carrying people) is it worth paying more for the displacement vessel's major drawback, restricted speed.

As we have already seen,[1] ships have been getting slower and it seems unlikely that there will be a major increase in ship speeds, at least as long as diesel oil remains relatively expensive, but not so expensive as to justify investment in alternative fuels. It seems unlikely that nuclear power will be generally adopted in merchantmen for cultural reasons as much as technical and economic. The main speed bracket for ships will probably, therefore, remain between 15 and 25 knots.

Although more novel types of vessel might appear, it seems unlikely that they will become a significant form of commercial sea use. Cargo-carrying submarines have never gone beyond the early German experiments of World War I, experiments that reflected Germany's inability to use the surface of the sea rather than any advantages of submerged transport. Nuclear power has made submersible merchantmen more practical technically, if not economically, and concepts have been suggested, especially for use in the Arctic as polar oilfields are discovered and suspected. But it seems unlikely that the high cost of construction and operation of commercial nuclear submarines will see any but the most limited construction of such vessels. Indeed, submarine barges drawn by ice-breaking surface tankers would seem to offer a much more cost-effective solution to the problems of arctic oil transport if pipelines are impossible or inadequate. Civilian submarines may proliferate in the future, but their role will probably be connected with the growing research into exploration of the ocean depths rather than as long distance transport.

Surface skimmers of various kinds will continue to be significant in short sea and riverine transportation, but, once more for economic reasons, it is difficult to foresee any use for them in commercial oceanic crossings. It is possible that some skimmer technology, e.g. hydrofoils or side-wall air cushion vehicles (surface effect ships) might be useful on short and medium-length routes with especially high value cargoes. A 15,000 ton 'Atlantic Ferry' has even been suggested, capable of carrying trucks (or 2,000 cars and their passengers) say from the USA to Europe, cruising at 130 knots! Power, however, would be a problem for such a vessel, as nuclear energy, with all its problems, would probably be essential. It is also hard to see who would produce the necessary investment. 'Wing in ground' (WIG) machines which use the ground effect to skim across the waves also might have greater potential for fast oceanic crossings. Gas turbines might be used for propulsion, although fuel storage would be a problem. Again, initial investment could well prove an even more insuperable difficulty.[2]

Indeed, all novel forms of sea transport have the overwhelming problem that there are cheaper and more established alternative means. Only if states wish to ignore economic factors for political, including strategic, reasons can this financial strait-jacket be overcome. It is well known that the Soviet Union, given its socio-economic system, has been able to put such factors high in its priorities for the impressive build-up of its mercantile marine. The United States, however, provides a more interesting example of the interplay of 'market' forces and politico-strategic factors. Now 'Sealift' has been officially defined as a naval role it is just possible that non-commercial considerations may come more into play in the development of fast vessels to carry supplies and fuel. Multi-hull vessels might lend themselves to such tasks, such as catamarans or even SWATH (small waterplane twin-hull) designs. The latter, while much less economic than conventional ships at normal operating speeds, offer less resistance at speeds above 37 or so knots.[3] They might, therefore, offer possibilities for ultra-fast cargo transports. The investment implications, however, are so enormous as to make it difficult to foresee the circumstances in which such ships would be built.

For the foreseeable future, therefore, naval warfare is going to continue to revolve around relatively slow ships. This means that platforms will be required of similar endurance and performance, i.e. military surface ships. The inherent load-carrying advantages of the surface ship also make it an excellent military platform in its own right. The surface ship's endurance, as Admiral Hill called it, its capacity for 'being there' in a wide range of conditions, give it an unrivalled capacity to maintain a patrol when a military or constabulary presence is required.[4] The surface ship also has unrivalled capacity to operate in three dimensions, and to communicate with other platforms.[5] It can even act as a mobile platform for other platforms, notably aircraft. Navies still cannot do without surface ships and will not be able to for a long time, if ever.

Surface warships

The term 'warship' covers a very wide range from the small patrol boat to the huge super-carrier. Naval planners must choose which are necessary for their strategic and operational requirements. For navies whose main role is constabulary, relatively small but high endurance vessels with a very limited armament and probably some helicopter capability are the best solution. There are a range of such vessels around the world's navies and coastguards today as the constabulary responsibilities of nations at sea have expanded. Archetypal vessels are the 1,150 ton Icelandic 'gunboats' of the Aegir class, diesel powered with a small 22-man crew and a speed of 20 knots. Some coastguards need much larger vessels (in the 3,000 ton bracket) for higher endurance, e.g. the US Coastguard and Japanese Maritime Safety Agency. For states with very short range responsibilities, e.g. the Bahamas, a 100-ton patrol vessel will do.

Platforms

Armament need only be very limited, the Icelandic vessels carry ninety-year-old Danish six-pounders. Toughness, seakeeping and manoeuvring characteristics are much more important for such vessels even in clashes with quite major naval powers over fisheries, as the British found out to their cost in the Cod Wars of the 1970s. Such patrol vessels, however, stand little chance against more aggressive opponents, as the Bahamians found out in 1980 when their patrol vessel *Flamingo* was sunk by Cuban Mig-21 fighters. If patrol duties move into shallows or shoals then hovercraft offer some advantages as patrol craft, combining speed and ubiquity with a certain level of endurance and an ability to land as well as just inspect.[6]

Many more combatant navies use their smaller MCM vessels for constabulary duties, as the characteristics of an MCM vessel are similar to those of a specialized patrol craft. The British coastal MCM vessels, built to counter the Soviet mining threat of the 1950s, found their major role in policing and constabulary duties all over the world. The problems here are that the increasing sophistication of MCM vessels and the need for intensive training in their primary role can make diversion of these valuable assets to other tasks difficult. A contemporary coastal mine hunter, with its computer-assisted action information system, sophisticated sonar, acoustic magnetic and wire-sweeping equipment, and guided remote-controlled mine locators, is an expensive warship, not available in large quantities. For this reason the United Kingdom, for example, has turned to simpler, more specialized, vessels for new MCM construction. Trawlers can also be pressed into service in emergencies for certain simpler MCM duties, e.g. the sweeping of deep-laid ASW mines. Given the need to keep signatures at a minimum and the coastal nature of the work, minesweepers are seldom much over 1,000 tons.

Serious fighting against other surface craft or aircraft is not the function of any of the above vessels. For this duty much heavier armament and high speed is required, at a cost in endurance. The world is now full of fast-attack craft armed with guns, missiles, and torpedoes for use against other fast-attack vessels as well as larger ships. They complement aircraft in coastal defence and operations in restricted waters, being especially useful at night. They can also project power over relatively short distances, especially as their size and armament increase, as the Israelis have shown in the coastal bombardments carried out by their 415 ton Reshef class boats. The Israelis have indeed become masters of fast-attack craft operations, even giving some of their latest vessels helicopters for over the horizon targeting of long-range Harpoon missiles. Fast-attack craft

The changing shape of naval war

can be very heavily armed indeed, e.g. a Reshef can carry a 76 mm gun, a Vulcan Phalanx close range weapon system, 2 × 20 mm guns, four to six 40 km range Gabriel missiles, and two to four longer range Harpoon missiles. Hydrofoil technology also lends itself to the fast-attack role, especially for the carriage of missiles and torpedoes. Such vessels are produced by a number of countries, including China, whose navy deploys hundreds of fast-attack craft of all types.[7]

The 'military sex appeal' of fast-attack vessels is enormous and they make junior officers' mouths water. They will certainly remain a most important part of the naval scene, making operations in coastal waters dangerous even for more powerful navies. The attention paid by the Americans to the Libyan fast-attack craft in the Gulf of Sidra confrontation was noteworthy, even to the extent of a jumpy USS *Yorktown* firing a Harpoon missile at a dubious contact, and sinking it without even being sure of what it had hit.[8] The Libyan fast-attack craft, built in France and deploying western Otomat surface-to-surface missiles, were probably the most formidable anti-shipping systems available to the defenders of the Libyan leader's claims. But aircraft equipped with the same missile might have been even more useful. Moreover, despite an undoubted ability to conduct long ocean transits, even quite large fast-attack craft in the 500 ton bracket are at the margins of seaworthiness for sustained and efficient oceanic operations. Moreover, their size prevents them from carrying the suite of sensor and control systems essential for fully effective ASW and AAW duties.

Only larger surface combatants can cope with these demands, the degree of size and sophistication reflecting the interaction of the duties that the ship is likely to be called upon to perform, the range at which it is expected to perform them, the weather conditions in which it is expected to operate, the threat it is likely to face, its likely uses, and last, but not least, the amount of money the national finance ministry can be persuaded to make available. Obviously, different nations can come up with different answers and will continue to do so. Sometimes the same navy comes up with a complementary set of answers, a hi/low 'mix' as Admiral Zumwalt called it in the 1970s, reflecting the balance of capabilities it perceives it requires.[9] It must also be stated that all these rational factors are affected by the nature of warships as tokens of national prestige. Sometimes this dictates a level of capability (and sometimes even an existence) that a more rational decision might have thought excessive.

If we look at the types of surface combatants being built for the world's navies we see an interesting array of corvettes, frigates, destroyers, and cruisers of different shapes and sizes. First, some words are necessary on nomenclature. For some time now, navies and naval analysts have been struggling unsuccessfully to come up with an agreed nomenclature to cover the surface units of the three-dimensional world of naval warfare that began in World War II. The old battleship/cruiser/destroyer distinction born at the turn of this century, a rationale embedded in the surface warfare of the gun and the torpedo, was already breaking down even before the World Wars saw the development of ASW and, later, AAW escorts for merchantmen which could not be fitted into the standard framework. Mercantile escorts were called attractive and often inappropriate names culled from the days of sail: first sloops, later corvettes, and then, as the latter got larger, frigates. Destroyers got larger too as their builders tried to acquire more margins of advantage over fellow navies' destroyers; indeed, they began to approach the size of the cruisers of old, especially as they acquired primarily ASW and AAW roles to deal with the new threats to the fleets. The only difference between destroyer and frigate became the higher speed required for operation with the new carrier battle groups, and as slow mercantile escorts increased in speed themselves to cope with newer submarines, confusion became inevitable. Different navies used the same word for a totally different type of ship, e.g. the Americans and French chose 'frigate' for their large destroyer/small cruiser category, a rather different ship from the British 'frigate', a less capable 'sub-destroyer' primarily mercantile escort category born in World War II. The USSR ignored the problem and began a new system of 'large' and 'small anti-submarine ships'. Some countries almost deliberately produced misleading 'small' titles for new construction, in order to ease them through the bureaucratic process, notably the British with their County class destroyers and Invincible class cruisers.[10]

The tendency of Western analysts to call Soviet 'large anti-submarine ships' cruisers, and the demise of the traditional large cruisers of the Dreadnought era, led the US Navy to reassess its nomenclature and come up with perhaps the best solution yet to the problem. In a major reclassification in 1975 it sensibly redesignated most of its frigates as cruisers and the rest as destroyers. It joined the rest of NATO in classifying its 'destroyer escorts' as 'frigates'. Yet even this system is now beginning to go fuzzy around the edges as the US Navy plans to build destroyers that will be as big as, and more capable in every way than, most current cruisers. Attempts to produce a comprehensive and widely acceptable system of nomenclature seem doomed to failure given the various customs of the world's navies. However, if the terms 'frigate', 'destroyer', etc. are to be saved, they need some content if they are to have any meaning at all. The following is offered as one solution which will be followed for the rest of this chapter; it will be noted that displacements overlap between the various types.

Cruiser: Large combatant with a comprehensive range of capabilities including area air defence: usually about 7,000 tons in standard displacement to give sufficient room for all required systems.

Destroyer: Medium-sized combatant of between 2,750 and 7,000 tons with the speed and capabilities of the most demanding combat operations, including participation in fast carrier battle groups. Large 'destroyers' differ from 'cruisers' of equivalent size in that they lack comprehensive capability, e.g. no area defence AAW system. Large 'frigates' usually differ from small 'destroyers' primarily in terms of lower speed and more austere engineering arrangements, e.g. single as opposed to twin screw.

Frigate: Combatant of about 1,750 to 3,000 tons usually optimized for ASW but with general purpose capability; essentially intended for the escort of non-combatant shipping, although useful for patrol and limited offensive operations; capable of ocean-going transits and tasks; usually air capable.

Corvette: Small sea-going combatant of 500 to 1,750 tons of limited endurance and capability; best employed in coastal escort and patrol work and relatively low-level operations, although with some potential for higher endurance operations in the absence of more capable units; not usually air capable.

Only the largest navies, effectively the superpowers, can afford cruisers which combine all three capabilities, AAW, ASW, and ASUW, in adequate proportions. The Soviet Union, only just joining the carrier business, has produced instead latter-day 'battle cruisers'. These interesting Kiev class 'heavy aircraft carrying cruisers' of 36,000 tons could lay claim to being the most comprehensively equipped warships in the world. They carry a mix of helicopters, missiles and rocket launchers for ASW, VTOL fighters, missiles, and guns for air defence (and limited ASUW) and long-range missiles for ASUW. They are clearly intended to form a centrepiece of any grouping of Soviet surface combatants. Equally impressive, if not more so to some eyes, are the four 24,000 ton Kirov class 'missile cruisers' with better long-range missile capabilities for both ASUW and AAUW than the Kievs', but only three helicopters and a generally more limited ASW potential. These would seem to be primarily fighting mobile command platforms that provide a capacity for sustained oceanic sea-control operations if required. Although there is evidence that these large vessels are something of an embarrassment to the post-Gorshkov navy, all eight of these ships will remain in service for some time as status symbols as well as instruments of Soviet naval power.

In the absence of carriers the Soviets naturally give a greater emphasis than the West to anti-surface warfare in the capabilities of their larger vessels. An impressive example of a modern ASUW missile cruiser of more moderate dimensions is the 10,000 ton Slava class with serried ranks of SS-N-12 anti-ship, long-range cruise missiles. It is protected by the Soviets' latest long-range surface-to-air missile (SAM) systems, and with torpedoes and rocket launchers to give adequate ASW protection (although the helicopter is primarily for surface-to-surface missile (SSM) targeting). A ship of this class will provide a potent and visible anti-carrier warfare capability of some sustainability: 130 mm guns even allow the means for some 'gunboat diplomacy'. Other cruisers, like the older Karas, emphasize ASW with missiles, torpedoes, and helicopters.

American cruisers reflect their primary role of providing anti-air defence to carrier battle groups. All carry long-range air defence missile systems as primary armament and the latest ships are the first to carry the ultra sophisticated Aegis system. All can also participate in ASW, although only to a limited extent in some ships due to lack of helicopters; even if a hangar is fitted an aircraft may not be. Some ships have sacrificed helicopters for cruise missiles but the Aegis class ships are fully helicopter-capable. SSMs have allowed the US surface sailors to enhance the capabilities of their vessels in ASUW, and both Harpoon and Tomahawk missiles are being fitted to the US Navy's cruiser force. The new vertical launch system will allow a comprehensive mix of AAW ASUW, and ASW missiles to be carried, although one suspects AAW missiles will predominate. It will be difficult, however, to stop the United States increasingly thinking of its cruiser fleet as a force of self-sufficient ships capable, like the cruisers of old, of a wide range of duties.

In the 1970s the United States thought of producing larger, more impressive ships still: nuclear powered 'strike cruisers' along the lines of Soviet vessels, albeit rather smaller. It was hoped that Aegis would give sufficient self-defence capability to operate independently. The ships would also be armed with SSMs and an 8 inch gun and would also be armoured.[12] Cost, and the supremacy of the air lobby, helped defeat the concept. Another factor was the reactivation of fast battleships of the Iowa class. These heavily armed and armoured vessels were built during World War II, with the high speed of cruisers and the range to operate with carrier task forces. They now provide large and impressive centrepieces for surface action groups, supplemented by a screen of cruisers and destroyers. They can be

The changing shape of naval war

given SSMs capable of dealing with targets far beyond the range of their guns, but not only expense advises against turning them into truly modern warships. Their invulnerability depends to a considerable degree on their ammunition and other flammables being kept deep inside the ship behind armour. An aircraft hangar and/or large topside missile magazine, might well compromise this kind of protection, probably to a decisive degree. As they stand, battleships may well remain for some time useful as coastal bombardment and 'firework display' vessels; they will also have useful long-range capabilities at night and in poor weather, especially in conditions of electronic silence. They will also remain useful symbols of naval strength. There is perhaps still a role for such symbols but one hopes that the battleships' active service will not ruin them as the historical relics they essentially are. The US Navy's surface sailors are already pitching for missile-armed battleship replacements which may possibly see the strike cruiser resurrected in a new, more powerful form. Whether such ships would be built, however, remains a very open question.[13]

Other old warships that may well see more service are the surviving large gun-armed cruisers. The United States retains in reserve two huge 17,000 ton Salem class ships which it might be premature to write off. Latin America, however, is the traditional home for these vessels with their imposing appearance and considerable surface bombardment potential. Peru has spent considerable sums (that she has had problems raising) on modernizing her two 10,000 ton, 6 inch gun ships purchased from The Netherlands. One has been modified to carry missiles. Chile, which still retains an old US-built Brooklyn class ship half a century old, is talking of converting her into some kind of aircraft carrier. Chile, however, like Pakistan, is 'trading down' to purchase very large destroyers as the 'light cruisers' to carry her into the next century. She has acquired no less than four of the British County class destroyers of over 5,000 tons, complete with a stock of Seaslug SAMs. The latter are probably nowadays more useful in a surface-to-surface mode (and were used for shore bombardment in the Falklands War). They add, therefore, to the ship's Exocet SSMs and 4.5 inch guns and the relatively modern action information suite to produce quite a capable modern light cruiser for the old fashioned South American naval environment. One ship, which had already lost its Seaslug, has been fitted with improved helicopter facilities and this is the near-term future for half the class.

Modern destroyers are, indeed, hard to distinguish from small cruisers, the key perhaps being increased role specialization. The superpowers deploy large and impressive examples of the 'destroyer'

Platforms

type, the different operational priorities of the two navies affecting the types of ship they will take into the 1990s and beyond. The Soviet emphasis on more 'offensive' uses of groups of surface ships in the ASUW and ASW roles, rather than the screening of higher value units, is still reflected in the new generation of 6,000–7,000 ton destroyers of the Sovremenny and Udaloy classes. Both are optimized for one form of warfare only, and neither has anything much beyond a point defence anti-air capability. Effectively the Udaloy is a smaller platform for the ASW capability of a Kirov (medium/low frequency, hull-mounted and variable depth sonars; Helix ASW helicopters; SS-N-14 missiles; and short-range rocket launchers), and the Sovremenny, a highly specialized ASUW vessel, with 130 mm guns and modern SS-N-22 100 km range sea-skimming missiles.

Again, USN ships as 'task force escorts' show a considerable difference in emphasis, with greatly enhanced AAW potential in most ships. The new DDG51 Arleigh Burke class vessels are in some way 'cruisers' as they will carry Aegis in a slightly reduced form: their lack of a helicopter hangar, however, denies them the superior classification. More simple AAW missile systems, although of medium rather than long range, will remain deployed for some time in older US destroyers. The large and fast Spruance class ships provide fast, more specialized ASW task force escorts. Again the tendency to fit ASUW potential is noticeable, with Harpoon in service and Tomahawk due to appear as ships are refitted with vertical launch tubes. In terms of general capability, in fact, the US destroyer force of the future will, to some extent as now, be distinguished from the larger cruisers purely by size. They will, however, all lack something in comparison to their larger brethren be it in terms of AAW (e.g. Aegis) or ASW (no proper helicopter facilities).

Both Britain and France deploy slightly smaller and cheaper destroyers than the superpowers. This makes task specialization even more essential. The British solution has been to build the Type 42 destroyer carrying the Sea Dart system for area AAW and to procure the Type 22 frigates (with low frequency, hull-mounted sonar, towed arrays in some ships, and better helicopter facilities). The Type 22 has been developed into a more general purpose vessel in its final version by enhancing the ASUW potential, and these four fine ships, *Cornwall*, *Campbeltown*, *Cumberland*, and *Chatham* will be important core units of the Royal Navy of the next twenty years. Sea Dart is an excellent long-range missile but the lack of point defence capability in the Type 42s caused serious problems in the Falklands. The Royal Navy had to abandon plans in 1981 for more impressive

The changing shape of naval war

Type 43 AAW ships which would have carried a mixed area/point defence missile system.[14] The Type 42s are now belatedly receiving Vulcan Phalanx CIWS that will allow them to cope better with high-threat situations. Their replacements look like being air defence variants of the NATO 'frigate' fitted with some kind of Aegis type air defence system procured under international auspices.

These ships will be similar in size to the 3,500 ton Duke class frigates of Type 23 now under construction. Much criticized for cost escalation, in truth these ships possess probably the minimum 'fleet destroyer' capability required for the Royal Navy's role as an integral part of the NATO striking fleet and 'first division' superpower naval warfare. With very long-range 2031 towed array sonar as their primary sensor and a hull-mounted, 2050 advanced active sonar set, the Type 23s will carry a large search/strike helicopter to prosecute sonar contacts, torpedo tubes for close range ASW, a 4.5 inch gun, and Harpoon missiles for ASUW and vertically launched Sea Wolf missiles for self-defence. Their advanced CODLAG (combined diesel electric and gas turbine) power plant makes them relatively slow (28 knots) but very quiet, an essential for effective towed array operations.[15]

France also follows the specialized route with AAW and ASW variants of its 3,800 ton C70 class escort (the C stands for 'corvette', a logical but confusing extension of France's previous 'frigate' category; however, the ships carry NATO 'destroyer' pennant numbers). The capabilities of these ships are not as great as those of their British counterparts, the Crotale point defence systems of the ASW ships are much less capable than Sea Wolf, and the American Standard medium-range system used in the AAW ships is not so long-ranged as Sea Dart. Crotale is, however, being upgraded to cope with the missile threat and to increase its range. The ASW ships have low-frequency, hull-mounted and variable depth sonars, and towed arrays are beginning to be used. The limited air defence potential reflects the French Navy's emphasis on operations outside the NATO area against more limited threats, as well as budgetary constraints caused by the nuclear emphasis of her strategic policy: the Standard missile fits in the two AAW ships are from older destroyers. Older, larger frigates also remain: the two 5,000 ton Suffrans with French-built long-range Masurca AAW missiles used primarily to escort France's operational carrier, and the three 4,600 ton F67s optimized for ASW with Lynx small search/strike helicopters and the Malafon ASW missile (also carried in the Suffrens). These ships will be in service until 1998–2003.

Pulling ahead of the European 'second rank' navies in the

destroyer category, and in terms of the individual quality of her units, is Japan. One must expect the future to see her deploying perhaps the most impressive surface combatants outside the superpower navies. The relatively limited reach so far required for her forces, and the physically small stature of her personnel, have already allowed her to produce some outwardly very impressive destroyers, four near cruisers of about 5,000 tons, which were some of the first to carry large search/strike helicopters as opposed to just weapons delivery machines. Japan plans, when funds are available, to procure full Aegis capability for its new generation of 6,000 ton destroyers. She has just built a pair of 4,450 ton Hatakaze destroyers optimized for AAW with SM-1MR missiles; guns and missiles provide supplementary ASW and ASUW capabilities, but there is no helicopter. Further down the Japanese destroyer scale are ASW ships in the 3,000 ton bracket, a dozen Hatsuyuki class ships recently completed, and eight slightly larger Asagiris are entering service between 1988 and 1991. All these fast and rakish ships, powered by Rolls Royce gas turbines, combine a low-frequency, hull-mounted sonar, large search/strike helicopter, ASROC missiles and torpedoes for ASW, a 76 mm gun and Harpoon missiles, and Sea Sparrow and Vulcan Phalanx for AAW self-defence. The Asagiris also carry a towed array, and the older ships are to be fitted with this vital addition to their capabilities.

NATO's major navies deploy destroyers of various shapes and sizes. Canada, Italy, and the Netherlands have modern ships in the 3,500 ton bracket that will run on for some time. Canada's current destroyers are platforms for long-range search/strike helicopters intended for Atlantic operations. Six new units are planned to enter service between 1989 and 1992. Italy has four air defence guided missile destroyers of 3,200–3,600 tons with two 4,400 ton ships projected. The Dutch have two new 3,665 ton frigates that have capabilities in all the dimensions of naval warfare, medium and short range AAW, missile defences, helicopter, ASW torpedo tubes, SSMs, and 120 mm guns. They provide excellent flagships for groups of smaller escorts, but even the Dutch could not afford one for each of their escort groups and have had to produce SAM-equipped versions of their 3,000 ton Kortenaer class frigates as substitutes, minus a helicopter capability. The standard Kortenaers, ten of which are in service, are primarily ASW vessels with a medium frequency, hull-mounted sonar, provision for two Lynx helicopters, and torpedo tubes. Equipment includes a 76 mm gun, Harpoon, SSMs, and point defence SAMs plus a Goalkeeper close-in weapons system complete the weapons suite. Towed array, currently fitted to two older frigates, may be fitted later, and this is also due to be fitted to the new, slightly larger Karel Doormans that will also carry a large Sea King type search/strike helicopter.

Related to the Kortenaers are the German Meko frigates of the Bremen class, which are powered by diesels as opposed to the gas turbines in the Dutch ships. The Germans plan to procure an improved Type 123 frigate to replace their old gun-armed destroyers in the early 1990s, while they are currently hoping the NATO AAW frigate will eventually replace their current guided missile AAW destroyers. The Kortenaer/Meko design has shown some potential for export with the former being purchased by Greece and the latter, in heavily modified form, by Argentina with an armament optimized for ASUW (5 inch gun, Exocet missiles, 4 × twin 40 mm AA guns) and gas turbine engines of, remarkably, British manufacture.

As we have already seen, large general purpose ships in the destroyer category have the ability to be the 'poor man's cruiser', allowing smaller navies to deploy surface combatants of significant size and prestige and some military utility in less sophisticated environments against less difficult opponents. An interesting example is the new Romanian 5,000–6,000 ton 'battle cruiser', effectively an ASUW vessel with Soviet type SS-N-2C missiles, 76 mm and 30 mm guns, 533 mm torpedo tubes, and a helicopter hangar. Along the same anti-surface lines, although rather smaller, there are China's fifteen 3,250 ton Luta class ships, commissioned between 1971 and 1985, which grafted Chinese copies of the SS-N-2 onto a destroyer hull of Soviet/Italian 1930s conception. India has been able to take an alternative path, purchasing from the USSR six newly-built versions of the 1960s vintage 4,000 ton Kashin class, which at least have some SAMs as well as SSMs. As a domestically produced supplement the Indians have enlarged the Leander class frigate design into the 4,000 ton full load displacement ASW frigate *Godavari*, actually an impressive destroyer, with the ability to operate two helicopters. Three of these ships are already in service and the design is being enlarged into the 5,000–6,000 ton gas turbine powered Project 15 frigate, the first of which was laid down in 1987. Pakistan showed an interest in purchasing British Type 23s to provide her with something approaching equivalent capability, but for the time being has settled for old, former Royal Navy and US Navy frigates to supplement her second-hand American destroyers.

One of the most interesting features of the next decade or two will be the extent of the destroyer development of the Chinese and Indian surface navies. The Indians seem to have abandoned their plans to buy Soviet Kresta IIs, but instead hope to purchase six to eight modern destroyers of the Udaloy and/or Sovremenny classes from the USSR, as well as build six of her own Project 15 destroyers in the 1990s. The Chinese are modernizing the Luta to make up for its manifold deficiencies. Vulcan Phalanx CIWS will provide last ditch air defence, while American ASW torpedoes and, in some ships, a helicopter, will give ASW potential. The Chinese have so far failed, for technical reasons, to produce an adequate SAM of their own and, for economic reasons, to find a foreign substitute. But a new SAM system seems to be under development now. A slightly enlarged, modernized Luta design has been exhibited, apparently with exports in mind. It mounts an American 5 inch gun, eight modern surface-to-surface missiles (American or Chinese), an unidentified SAM system, Vulcan Phalanx CIWS, ASW torpedoes, and a helicopter. China will need to replace her own destroyer fleet with large numbers of this kind of ship before she possesses a true 'blue water' naval capability. This will take well into the next century.

Many of the smaller modern destroyers in the world today are often classified as 'frigates' by their owners. Where to draw the line between the large frigate and the small destroyer is very much a matter of national taste. Nigeria deploys an identical ship to Argentina's German built Meko 360s: in the former navy she is a 'frigate', in the latter they are 'destroyers'. For the purposes of this analysis, however, the criterion of suitability for main fleet open ocean operations dictates a lower destroyer cut-off point of about 2,750 tons standard displacement, at least for twin-screw ships without some significant limitation on their capability. Ships marginally larger than this may remain 'frigates' due to some significant limitation or other, e.g. single-screw propulsion.

The largest frigates, rather than destroyers masquerading as such, are those deployed by the superpowers. American frigates are in the 2,750–3,000 ton category but only have single-screw machinery which restricts their performance. They are intended to screen underway replenishment groups and military convoys rather than act as fleet escorts (although they are often used in the latter role). The latest US frigates, the gas turbine powered FFG7s, have comprehensive sonar equipment, multi-purpose missile launchers, large helicopter hangars, and 76 mm guns, and have been exported to Australia and Spain as well as procured in large numbers (51 units) by the US Navy itself. Having already procured 46 very large 3,000 ton Knox class steam powered ASW frigates between 1969 and 1974, the US Navy, paradoxically for an essentially power projection force, now has the largest fleet of convoy escorts in the world.[16]

The changing shape of naval war

The Soviet approach to the larger frigate is also a 3,000 ton vessel with four long-range ASW missiles, shorter-range ASW rockets and torpedoes, and 76.2 mm, or 100 mm automatic guns. The Krivak class ships, which might be destroyers but for their limited potential, were subject to a great deal of unwarranted admiration when they appeared in the early 1970s. A new variant has now appeared carrying a helicopter instead of missiles. This seems to be a modification for coastal patrol operations as the ships are operated by the border guards rather than the Navy. Paradoxically, however, the escort capacity of these vessels is higher than their more combatant predecessors, which are not suitable for much more than patrol or picket duty, as implied by their new post 1977–78 classification of *storozhevoy korabl* (patrol ship).

More classical frigates come in two size bands, one of about 2,500–2,750 standard displacement, and the other between 1,750 and 2,250 tons. The archetype of the larger class is the classic Leander, developed by the British from the Type 12 fast convoy escort frigate built with all the lessons of the World War II Battle of the Atlantic in mind, the campaign that spawned the original modern 'frigates'. At the end of the 1950s, the British developed the Type 12 concept into a 'general purpose' vessel that was used as a fleet ASW escort and as a colonial sloop for presence and peace-keeping duties: the first examples were commissioned in 1963 and no less than seventy-two ships of the class were built in three countries. The Royal Navy received twenty-six, many of which are still in service, some in much modified form to enhance their capability as 'ersatz destroyers'. The 'Leander package' – displacement of about 2,500 tons, medium gun armament, and/or SSMs, point air defence system, relatively powerful hull-mounted sonar and weapons delivery helicopter, sufficient radar fit for fighter control, and aerial and surface surveillance – is the shape of the modern frigate, a ship of great operational flexibility.[17] It can be seen in the British Type 21s built in the 1970s, and the Italian Lupos built for both domestic consumption and for export in the 1980s. The latter provide an interesting example of the impact of operational demands on ship size. Originally conceived as a 2,200 ton convoy escort, an extra 300 tons were added to make the frigate into more of a small destroyer for fleet operations with improved hangar, variable depth sonar, and improved accommodation. A large frigate like this, rather than a full-scale destroyer, is adequate for higher-level operations over the limited distances and in the shallow waters of the Mediterranean.

British naval shipbuilders have been constrained by their domestic navy being more in the market for small destroyers than 'real' frigates. Various private venture designs are currently on offer, however. Vosper Thorneycroft have taken the Leander concept and modernized it into a 2,800 tonne, general purpose frigate with diesel or combined diesel and gas turbine propulsion with electric slow-speed drive. The ship can engage submarines with the helicopter or ship-launched torpedoes; surface targets can be engaged with Harpoon missiles, 5 inch guns or SSMs carried by its helicopter; and it can defend itself from the air with Sea Sparrow missiles, and its gun and Phalanx close-in weapons systems.[18] The Greek Navy showed some interest in procuring a batch of these vessels but instead returned to a requirement for large Dutch destroyer sized 'frigates'. The 2,000 ton frigate bracket is, however, the one usually preferred by export customers and it is a highly competitive market. The small Italian Lupo design has been purchased by several foreign customers, Iraq, Peru and, Venezuela. France sold four 2,000 ton F2000S ships to Saudi Arabia in the mid-1980s. The latter are general purpose vessels with a small helicopter, ASW torpedo tubes, 3.9 inch and 40 mm guns, Crotale short-range SAMs, and Otomat medium-range SSMs. The West Germans are building slightly larger MEKO 200 frigates for Turkey, and Greece has been tempted by these ships as its smaller alternative to the larger Dutch vessels.[19]

The 1980s marked something of a slump in the warship market due to difficult economic conditions in the developing world.[20] This made second-hand warships attractive once more. Indonesia, with a requirement for relatively long-distance cruising ability due to her paucity of bases in the large archipelago, purchased 1960s vintage frigates from Britain and the Netherlands. Such vessels are often recently modernized by their previous owners, and are capable of some years of extra service: their withdrawal from their home navies is as much due to financial pressures as technological obsolescence.[21] Older ships like the 1950s vintage Dutch gun-armed destroyers sold to Peru in 1980–82, or the old American destroyers of wartime construction distributed in large numbers around the world, can even now be attractive frigate substitutes for navies whose primary ASUW requirement reflects the naval priorities of an earlier age. Such vessels may, indeed, have the 'military sex appeal' that a more modern ship lacks. Although negotiating for a batch of domestically built FFG7s, Taiwan currently relies completely on ex-US destroyers of various wartime classes for its major surface combatants in the 2,000 ton category. She deploys no less than twenty-four such ships, the most recent transfers having taken place in 1983. Several of Taiwan's ships, like the later transfers to other navies, had been modernized by the US Navy with ASROC ASW missiles

The changing shape of naval war

and a hangar for a small helicopter. The Taiwanese have continued the modernization process, fitting their locally produced 40 km range Hsuing Feng version of the Israeli Gabriel SSM, as well as additional light guns and the simple Sea Chapparal missiles to provide point defence AAW. Sea Sparrow and Standard surface-to-air missiles are planned to convert ten ships into specialized AAW vessels.[22] It is a tribute to the quality of American wartime design and construction that these hulls will still be in service over half a century after their construction in the mid-1940s.

Such old hulls as these, and more modern frigates of similar size, give a navy the opportunity to carry out sustained operations at sea. This, however, may not always be necessary. If one is willing to accept serious limitations in seakeeping and military capability in one or more dimensions, then the corvette of 500–1,750 tons has some utility, even for superpowers. The USSR which still gives coast defence a very high priority, deploys a large number of rocket and torpedo ASW corvettes in the 500–1,000 ton class, as well as missile corvettes that are really large fast-attack craft. Both types of corvette are capable of limited sea duty, e.g. in the Mediterranean, but for those states who might need a little greater sustainability a larger 1,700 ton Koni class is built for export to friendly states such as Cuba, East Germany, and Yugoslavia. These ships have limited abilities above, on, and below the seas with ASW rocket launchers, light 76 mm guns, SA-N-4 point defence missiles and sometimes anti-ship missiles. Such large corvettes rather than small frigates, notably lacking in air facilities, provide their owners with a very limited, but possibly sufficient, ability to deploy a little force a little way from home, but not too far.

Second rank navies with foreign bases and a need for a basic presence in distant seas also see corvette-type vessels as attractive. With their 'out of area' priorities, the French deploy a crude 1,100 ton 'aviso' or 'sloop' with some capabilities in all three dimensions, which has also been exported. Their larger 1,750 ton predecessors were also little more than colonial gunboats, even to the extent of having ASW launchers that could double as shore bombardment weapons. The Iberian navies, with their colonial background, have found corvette vessels of this type attractive, but also find difficulties in breaking the habit, despite such ships' limited capabilities. A modern Spanish Descubierta 1,720 ton corvette can shoot at surface and air targets with guns and missiles, and engage submarines with torpedoes and rockets, but all at only the closest ranges. They can also carry thirty troops each which gives more than a clue to the type of duties at which they are best. Yet even nations such as Japan feel that for their shorter ranged and shallow water operations a

Platforms

small corvette type of ship with ASW rocket launchers (and Harpoon surface-to-surface missiles) is useful. This DE (destroyer escort) concept is being enlarged with ASROC in new construction to 1,900 tons, which may better qualify it for the 'frigate' title, despite its lack of a helicopter. With its rather similar Baltic requirements, Denmark has built similar, but rather smaller, British-designed vessels. The little 1,320 ships were designed as classic shallow-water ASW vessels intended to work in home waters with shore-based air support. A hull-mounted sonar and ASW torpedoes would provide the primary systems while ASUW potential came from Harpoon missiles and a 76 mm gun; Sea Sparrow would provide close AAW cover. In service the ASUW role seems to have come to dominate: the torpedoes have not been fitted and the only ASW weapons available are depth charges.

Helicopters can be added to the large corvettes in the 1,500 ton bracket, but the aircraft must be small and the operating conditions are always marginal. The FS1500 large corvettes/small 'frigates' built by West Germany for Malaysia and Colombia have helicopter facilities and the latter even have hangars. Hangars are being added to Argentina's similar-sized Meko 140 ships, but the latter vessels in particular seem very crowded and cramped. Indonesia has fitted a helicopter hangar to one of its Dutch built corvettes and the owners contemplate landing and taking off helicopters from the 670 ton corvettes purchased by Ecuador from Italy in the early 1980s. Middle Eastern navies contemplate helicopter operations from very small vessels. Indeed, Israel has even added a hangar to two of her 500 ton large fast-attack craft and plans the same for her new 1,000 ton corvettes proper. Such helicopter operations must be severely limited by weather and are probably more practical in the Mediterranean than the South Atlantic.[23]

Corvettes provide a suitable format for the newly industrialized countries to try their hand at building sophisticated warships. South Korea has produced a 1,600 ton Ulsan design for its own navy; they are basically ASUW ships, but more than a match technically for the crude 1,200 ton vessels built by North Korea. China has also long produced 'frigates' in this class, enlargements of the Soviet coastal escort design of the 1950s. China is enlarging this concept into the 2,000 ton class, and South Korea may well follow, both for domestic construction and export. The latter's greater access to Western technology gives its products a distinct technical edge, although at a price. Indeed, as South Korean labour costs increase its products might well begin to cost almost as much as European built vessels.[24]

Warship building outside the traditional naval shipyards of

Western Europe, the Soviet Union, and America may well increase. Already there are clear economic advantages in contributing as much as possible to the construction of one's own ships; doing so is considered an important way of creating and developing a shipbuilding industry. However, warship design is a specialized art and the design work still tends to be concentrated in the industrialized world, as does the provision of advanced weapon systems. This means that the domestic shipyard may receive little of the total cost paid by its government for the ship. Domestic construction can also have other disadvantages, e.g. sabotage. The optimal current compromise seems the co-operative technology transfer deal between developed and developing countries, with lead ships built in the originating yard and the rest by the purchaser. India, however, has an ingenious combination of Soviet and Western technologies. With the Godavari class destroyer, India shows what can be achieved by the Third World, albeit with some limited design help from foreign consultants. This seems the obvious and natural next step beyond building a ship completely to someone else's design.[25]

International co-operation may well also become more normal among the navies of the developed world. No navy, not even the richest, can afford as many surface combatants as it would like. Costs can be limited be engaging in fully international projects to spread costs over a number of countries. This latter technique also has the advantage of making the programme much more difficult to cancel or reduce. Given all the major European navies' problems of maintaining numbers as their older ships approach obsolescence it would seem that the 1990s might well finally see the adoption of a common 'NATO frigate' (actually more a small destroyer), perhaps with different weapons fits for different countries. There were some difficulties in getting all the nations to sign up for the project, but it is being pursued with more seriousness than earlier, similar, but in the end abortive, endeavours.

As for the basic form of future surface combatants, it seems likely that monohulls will predominate. There is still much room for development in the surface warship. The application of advanced technologies to monohulls can improve carrying capacity and combat ability for a given size of vessel. Vice-admiral Joseph Metcalf, as Deputy Chief of Naval Operations (Warfare) in the US Navy in 1986–7, supervised the setting up of machinery within the US Navy to study the impact of the 'Revolution at Sea' wrought by both the new weapons and sensor technology that has increased battle space 'up, out and down', and the control technology that allows highly automated systems. Metcalf's concept of the warship of the future is a long, low 'strike cruiser' with a speed of 31 knots and

Platforms

a ten decibel radar signature. Most of the ship would be composed of vertical launch missile cells. Gas turbines with superconducting electric drive would provide a compact power source for propulsion and, perhaps, beam weapons. Phased array radars mounted round the whale-like hull would be able to control missiles, which with command guidance, autopilots and active terminal homing, would not need illumination of the target from the moment of launch. A range of missiles would be carried for all tasks, ASW, ASUW as well as AAW. These ships would devote much more volume to weapons than has been customary in recent years. Metcalf has been willing to ask the difficult questions:

> . . . the matter of navigation and signal bridges . . . why [do] we have one on a modern warship[?] Both add space and weight. Bridges contribute to our ships' high radar cross section. Manning both bridges requires people who would be better used elsewhere in a ship's limited personnel allowance. The function and location of Navy ships' bridges are examples of policies based on history. Over the centuries, ship control has moved from the quarter deck on a sailing ship – located well aft, near the helmsman – to its present location in the bridge. The bridge's location high in ships is a throwback to the days when the captain watched the fall of shot. Today, he absorbs the output of countless electronic devices as he fights his ship from the combat information center (CIC) . . . . Could we co-locate ship control with the CIC, where the captain fights the ship and put both deep inside the ship?[26]

Like their immediate predecessors these futuristic surface 'flat iron' combatants might well be somewhat broader than classical warships. Contemporary warships emphasize beam in their designs as a result of the need to combine maximum topweight with the enhanced survivability gained from the use of heavier steel building materials. Both the new British Duke class frigates and the USN DDG51 have length/beam rations in the 7.6–7.7 range, compared to over 9 for British and American equivalents of the late 1970s. Some in Britain suggested that this should be further reduced to about 5, and a major controversy broke out between the advocates of 'long thin' and 'short fat' ships. Broad-beam vessels usually suffer from inherent problems of higher resistance and inferior seakeeping qualities. The supporters of the 'short fat' ship claimed, however, that their design overcame most of these difficulties and, indeed, reversed the balance of advantage. They claimed that their experiments, which included extensive tests with scale models, demonstrated that, contrary to the conventional wisdom, a considerable increase in beam is possible for any given speed

The changing shape of naval war

without increasing power per ton of displacement. The short fat ship had, they argued, a superior propulsive coefficient (effective horsepower divided by installed horsepower) compared to conventional vessels. The reasons for this were, however, unclear and seemed to be related to dynamic lift. However, such lift only gives significant advantages at high speeds that would require very large amounts of power to reach. As far as seakeeping is concerned, a 'short fat' hull, while less wet forwards in heavy seas, is still probably prone to rapid rolling, creating operational difficulties, notably the launching and recovery of helicopters, and making life unnecessarily unpleasant for the crew. Even the small patrol craft already built to the 'short fat' design have not, apparently, had entirely trouble-free careers in difficult sea conditions.

By far the weakest claim for the 'short fat' supporters was that a 'short fat' ship would be cheaper through its ability to carry more weapons for a given displacement. If other disadvantages of the design are ignored, it seems possible that a 'short fat' ship should be able to carry more equipment and, indeed, place its sensors at a higher and more effective level; but there is no way in which such a ship would be cheaper. Ship steel usually accounts for less than 25 per cent of the cost of a modern warship (in a British Type 23 it is 17 per cent). Propulsion represents about another quarter. Weapons and command and control systems account for over 55 per cent of the cost. If a 'short fat' ship can carry more systems than a 'long thin' one it must be more expensive, not less. Eventually the British Government was forced to commission an independent inquiry which compared enlarged versions of the original 'short fat' frigate design with the Type 23 frigate. It found that the former were, in fact, more costly than the Type 23, and that overall the more conventional design had better seakeeping. The inquiry, therefore, found 'no reason to disagree with MOD's preference for the Type 23 and it seems unlikely that major warships of "short fat" configuration will obtain general acceptance'.[27] Indeed, the trend could go the other way, with 'slender' vessels having advantages in certain applications, e.g. as a fast ocean escort for 'sprint and drift' submarine detection duties with subsidiary ASUW capabilities, or as an off-shore patrol vessel with good sprint capability. One less extreme 2,500 ton 'modern monohull' ASW/AAW ship is slightly longer in relation to beam compared to modern frigates and destroyers and has a maximum proposed speed of 41 knots with gas turbines and superconducting electric drive.[28]

Given the low relative ship steel costs the option of a deliberately 'enlarged ship' might be taken up. A larger ship has better speed and seakeeping qualities than a smaller one, and if the tendency to fill the spare space with expensive equipment is vigorously resisted, the result would be a cost-effective answer to the problem of maintaining the operational effectiveness of a given sensor and weapons suite at high speeds and in poor weather conditions. In some ways the current US Spruance class destroyer is such a vessel.[29]

Double-hulled vessels, catamarans, have certain advantages for specialized work, e.g. as small, fast patrol boats. Catamarans with their slender hulls are fast and economical of power, and seakeeping is also good up to a certain sea state. A planing catamaran is intrinsically more stable than a planing monohull, and foils placed between the hulls improve both seakeeping and fuel consumption. A German designed 32 ton Hyuscat-18 was built in Thailand in 1987 along the latter lines, armed with a Sea Vulcan rotary gun like a helicopter gunship. It is hoped to be useful for operations both in rough seas and shallow rivers. As for slower displacement catamarans, their lack of sensitivity to large amounts of topweight means that propulsion machinery can be mounted high up, away from underwater shocks and where it gives the least possible acoustic magnetic signature. This, together with shallow draught, makes for an excellent MCM vessel. The Australians have pioneered this concept with their Carrington minehunter now in service, and France is pursuing the idea with its *Batiment Anti-Mines Oceanique* due in the 1990s.[30]

Returning to monohulls, modern design gives planing craft some greater utility in more difficult sea conditions. It is possible that 'supercritical' planing hulls of the Sea Knife kind may be able to compete with the hydrofoil as a means of achieving high speeds in heavy seas.[31] For the moment, however, the hydrofoil still offers the best possibilities for high performance vessels required to operate in a wide range of sea conditions. The seakeeping conditions are such that ocean-going hydrofoils can be contemplated in the near term. As Michael Eames has argued: 'Hydrofoil development should proceed in the direction of simplification, cost reduction and improved range, even with some sacrifice in speed. This simplified hydrofoil will be attractive up to about 1,000 tons. In these smaller sizes the concept offers its unique capability of matching the seakeeping and exceeding the speed of major combatants.'[32]

Other types of skimming technology, such as hovercraft and wing in ground machines, have obvious potential in specialized roles, where a direct interface with the water is a drawback. Such duties as MCM, amphibious operations, and patrol work among sandbanks (e.g. in the Gulf) are obvious applications for hovercraft, and they are already used quite extensively for such duties, especially amphibious landings which promise a 'dramatic impact' operationally.[33] Hovercraft can be quite heavily armed: the British Hovercraft BH-7 is offered in fast-attack craft form with surface-to-surface missiles, point defence SAMs, and two 30 mm guns. An ASW version with dunking sonar is also available.[34]

Such craft are, however, only suitable for coastal duties. Of much higher endurance is the hovercraft's offspring, the 'surface effect ship' or SES. Edward Butler has summed up the advantages of SES: 'As compared to an aircushion vehicle (ACV), the SES hull, which pierces the water (hence non-amphibious) has less air leakage, better longitudinal stability, and an acceptable form of utilising water propulsion systems, which, at speeds to about 120 knots are more efficient than air propulsion.'[35] (Normal hovercraft use aircraft-like propellers). The Americans seemed set on utilizing SES technology to produce a '100 knot navy' in the 1970s. Two 100 ton prototypes were tested extensively but in 1979, an 80 knot, 3,000 ton prototype was cancelled only three weeks prior to its construction beginning. One problem was finding a mission for such a large SES with such a different operational envelope from the rest of the surface fleet.[36] Development continued in the 1980s with SES-200, a 200 ton trials vessel. The Coast Guard purchased three 110 ton Sea Bird vessels for coastal patrol, while the Navy ordered a class of 400 ton MSH mine countermeasures vessels for in-shore duties. The latter programme ended in fiasco when, after the first unit had been laid down, the design proved too vulnerable to shock. MSH-1 was abandoned in 1986, the SES design being replaced by a conventional Italian designed monohull.

Hardly more fortunate were plans to procure a 100 ton surface effect 'special warfare craft medium' (SWCM) for the US Navy's SEALS (sea-air-land special forces). 'Technical and scheduling problems' intervened and the contractor went into bankruptcy. The status of the SWCM programme has been described as 'tenuous', and again foreign monohull alternatives have been suggested.[37] Despite all these tribulations, consideration has been given to a larger Coast Guard cutter with a helicopter deck and a 600–800 ton Navy patrol craft for continental shelf ASW, surveillance operations 'associated with national interest and activity in hemispheric security',[38] and a fast replacement for current landing craft.

Norway has chosen the small SES catamarans for its new MCM programme and is considering the concept for a fast-attack craft. Given, however, the troubled development history of the SES concept so far, the application of the technology to fully ocean-going platforms is much more uncertain. Such surface-effects ships must (a) have the potential to be integrated tactically and logistically with existing fleet units, and (b) be an incremental development of current

Platforms

designs to minimize both technical risk and development costs. As it is few navies, if any other than the American Navy, have sufficient marginal expenditure to find such development attractive. Edward Butler, a leading exponent of SES technology, argues that the most desirable and practical large SES would be an escort/picket vessel in the 1,500–2,000 ton class. This would possess the seakeeping qualities of a larger ship and would enjoy the tactical advantages of being able to work in quiet waters listening on towed array sonar at normal fleet speed, but being able to work up to twice that speed to reposition itself in relation to the convoy or task force being protected. It would also be able to move quickly as a sensor or weapons platform as part of the outer AAW defence perimeter of either formation. Such a variant might carry Aegis and up to 80 missiles in vertical launch tubes. As ASUW derivative has also been suggested as a stealthy platform for high speed cruise missile attacks at ranges of up to 1,000 miles from the main formation. Such a vessel would be 'the least expensive missile attack mode that could be put together should this mission be needed. Its survivability should be high.'[39] Finally, a relatively small SES might be useful as a highly mobile aircraft platform given both the increasing use of aircraft and the ability of an SES to 'provide more flight deck area per displacement ton than any other' hull form.[40] Such SES ships could be made of one basic form interchangeable between missions.

A more suitable 'exotic' ship for the larger surface warship categories is the SWATH vessel. SWATH offers better performance in terms of the surface warship's main essential attributes, its capacity to operate for long periods in a wide spectrum of sea states and its capacity for effective presence close to the form of sea use being safeguarded. Compared to a monohull, SWATH offers greatly improved seakeeping and higher speeds in rough conditions. Rough seas act as a severe constraint on the operations of conventional warships. A 1983 US Navy study of its cruisers, destroyers, and frigates demonstrated that they could achieve full speed only up to low sea state 5 (moderate conditions with six-foot waves). In the North Atlantic in winter, frigates could only achieve full speed 30 per cent of the time, and destroyers and cruisers 55 per cent of the time: on a whole year basis the respective figures were 45 and 70 per cent. Low speed was not the only problem: helicopter operations, hull-mounted sonar searches, and at-sea replenishments were all deemed equally impossible by the USN beyond low sea state 5, conditions which occur almost 40 per cent of the time north of the equator. SWATH vessels would be much less limited, especially for air operations which would also be facilitated by the larger and higher deck area available.[41]

The changing shape of naval war

For ASW operations SWATH vessels provide excellent sonar platforms. They are perhaps the optimum towed array vessels, being able to move steadily in any direction in the widest possible range of sea conditions. The low deck motion, dry decks and inherent quietness of SWATH ships are also fundamental assets in this role. As platforms for hull-mounted sonars a SWATH ship's transducers would not be prone to coming out of the water: moreover, the twin cylindrical hulls would be excellent mountings for the wide aperture conformal sonar arrays that are expected in the future. SWATH's suitability for signature reduction, as well as improving offensive ASW potential, also enhances survivability, especially as a broad SWATH vessel is also less prone to catastrophic damage from a hit on one side of the waterline. The SWATH layout does have disadvantages, e.g. its inherently higher frictional drag leads to lower speeds in calm waters for a given power output and hence higher fuel consumption at normal cruising speeds, excessive draft, higher construction costs, and inherent difficulties in finding arrangements for weapons and machinery. Although steadiness is good, *stability*, the ability to return to normal after being disturbed, is poor. SWATH's vulnerability to flooding also seems 'excessive' to some.[42] On balance, however, SWATH's advantages seem substantial for ships designed to operate in rough conditions. Such vessels could be smaller and cheaper than conventional monohulls; advanced electric propulsion systems, coupled with contra-rotating propellors, might solve the fuel economy problem. The US Navy has ordered SWATH ocean surveillance ships (T-AGOS) for use in especially heavy sea conditions, and a SWATH frigate has been seriously considered. Studies demonstrated that on a calm water basis a SWATH FFG was 13 per cent more expensive than a monohull but that on a seakeeping basis she was 9 per cent cheaper. Moreover, 'incremental improvements in seakeeping are more cost effective using a SWATH hull form'.[43] As the United States is currently following a co-operative policy for FFG procurement and a low-risk monohull has already been chosen for this craft, the near-term prospect for a USN SWATH frigate seems low. Nevertheless, such a ship may well have been built somewhere by the end of the 1990s or in the early part of the next century, and the application of SWATH to small carriers or amphibious assault ships is a realistic possibility in the twenty-first century.[44]

More visionary proposals still have appeared to exploit SWATH to give surface units radically new shapes: Captain S.E. Veazey, ex Chairman of Naval Systems Engineering at Annapolis, has proposed a family of ellipsoid saucer shaped vessels, the optimum 'shape in which to distribute antennas, sensors and defensive and offensive

Platforms

weapons to obtain the most effective hemispheric view'.[45] Perhaps the greatest problem with such vessels is their unconventional 'science fiction' shape. This more than other considerations will probably tend to rule them out. Attracting scarce defence resources is perhaps going to be an insuperable problem to such radically new vessels, although they might evolve over the next century as SWATH vessels of more conventional layout become more common.

One recently produced analysis of the 'most promising options' for various warship designs has balanced the various factors for each hull form.[46] Table 6.1 summarizes the situation.

One important factor in future warship design will be the need to economize on manpower. This affects all countries as trained manpower will be a scarce and expensive resource, even to states with high rates of population growth. Modern ships are much less manpower intensive than their predecessors. A World War II destroyer, or even a 1960s vintage Leander class frigate, required 250–275 men as ship's company; the new Type 23s require just over 150. This may be taken further still. A British consortium (CAP Scientific Ltd. and YARD Ltd.) has produced a design study for an FCF-50 4,200 ton, 28 knot 'frigate' manned by only fifty people. The ship is heavily armed and carries a large helicopter. The very low ship's company was obtained by minimizing the at-sea maintenance requirement, within a sixty-day patrol cycle for the ship. On-board systems are fully automated and designed for the maximum availability and maintainability. Administration and support would be shore based with crews being rotated between patrols. The ship would be highly automated with data fusion as well as decision support at all command levels. Faults and damage would be diagnosed and repaired automatically, and duplication of systems and control centres would assist with damage control. The project shows what might be achieved in manpower savings in the near to medium term, at least for navies that do not need to operate over global distances and too far away from land.[47]

One factor in reducing manpower demands has been the adoption for warships of the internal combustion engine in its two main forms, the gas turbine and the reciprocating diesel. These are often combined in various ways to exploit the best characteristics of each type of engine to optimize fuel economy over a range of speeds. A diesel might provide cruising power while a gas turbine provides high speed capability. This is combined diesel or gas turbine (CODOG). A more complex arrangement with either diesel or gas turbine power, or a combination to obtain the highest power outputs, is called CODAG – combined diesel and gas turbine. Ships with only gas turbine plants usually have small cruising turbines (e.g. the

Table 6.1 Most promising future seaborne platform options

Ships' hull forms	Small aircraft carrier	Escort carrier	General purpose cruiser (ocean combat ship)	AA ocean escort	ASW ocean escort	Search & ASUW ocean escort	Ocean patrol vessel	Inshore patrol vessel	Assault landing craft	Amphibious assault ship	Mine countermeasures vessel
Conventional monohull			6,000–12,000t				1,500–3,000t			12,000–24,000t	< 800t
Slender monohull							800–1,500t				
Semiplaning monohull								200–400t	200–400t		
Enlarged monohull				6,000–12,000t							
Swath	12,000–24,000t	6,000–12,000t			3,000–6,000t						
Hydrofoil						400–800t					
Hovercraft									100–200t		100–200t
Surface effect ship					1,500–3,000t						

Rolls Royce Tyne) and large full-power turbines (e.g. the Rolls Royce Olympus). This is known as COGAG (combined gas and gas) if the two can be used in combination, or COGOG (gas or gas) if the two work separately. Gas turbines may well become more fuel efficient but they will remain expensive. It is significant that the Royal Navy, which has built some of the largest gas turbine powered warships in the world (notably the Invincible class carriers), is returning to diesel as its main cruising power plant for its new generation of frigates, although with gas turbine boost for high-speed dash (an installation called CODLAG). The new French Cassard class guided missile destroyer, on the other hand, is all diesel (CODAD). Without extra machinery, diesels are now able to produce adequate power levels for high service speeds and with their less technically demanding maintenance requirements will remain attractive, especially to less sophisticated navies. Dispensing with turbine air intakes and exhausts also helps save space in smaller vessels. Electric drive now offers excellent quietness, even to the inherently noisy diesels, and will be adopted more widely. Low-temperature conducting techniques offer future reductions in weight and volume together with a power storage facility for beam weapons or pulse radars. Electric drive might also allow propulsion to be put in external pods with advantages of redundancy and dynamic efficiency.[48]

Oil-fired external combustion steam turbines may be abandoned in new naval construction, although there are possibilities, especially in the Third World, for new remotely-controlled, high-technology, low-maintenance systems reversing this trend given the 'unreliability and high levels of maintenance with associated high operating costs' of gas turbines and higher-powered diesel installations. Fuel for the latter types has to be of high quality and specific consumption is also high.[49] Many countries would be willing to continue to pay this price for the flexibility of high-powered internal combustion engines but others may not. The Soviets may continue to employ oil-fired boost boilers in future nuclear-powered vessels, as they do in the current Kirov class ships. As for nuclear power itself, this looks like being restricted for the foreseeable future only to the largest warships, and these are almost all aircraft carriers. Even the United States, which in the 1960s and 1970s built limited numbers of nuclear-powered cruisers in the 9,000–10,000 ton bracket for carrier screening duties, seems to have little interest in building any more of the ultra-expensive and (for their role) somewhat over-sized units.

Nuclear propulsion brings us to aircraft carriers, the most impressive and powerful vessels of the present time and the near future. Their demise has often been prophesied, but the US building programme and their service life means that the twenty-first century will still see monster carriers of almost 100,000 tons full-load displacement steaming majestically across the world's oceans. Such has been the achievement of former US Secretary of the Navy John Lehman, a carrier aviator himself. Before achieving his office in the Reagan Administration, Dr Lehman wrote an academic study setting out the arguments in favour of these impressive behemoths.[50]

Size, he argued, gives significant economies of scale. Not only does it require less power per ton to drive a supercarrier at speed through the sea than a smaller vessel, but the volume of any container increases as the cube of the ratio of enlargement of its dimensions. The lower tendency to pitch and roll of larger ships and their greater seaworthiness means both lower rates of operational accidents and a narrower band of conditions in which aircraft cannot be operated for safety or physical reasons. Only a larger carrier can operate a comprehensive air wing of fighters, day and all-weather strike, airborne early warning, electronic warfare, ASW, and tanker aircraft. This is due to the big carrier's ability to operate a wide range of aircraft types, as well as aircraft in large numbers. With four catapults and large lifts and a large parking area, a supercarrier can also generate the very high sortie rate required in the vital early stages of operations in high-threat environments.

On the question of the apparent vulnerability of the large ship caused by its size and visibility he countered:

> It is one thing to put a hole in the deck of a carrier, it is another to damage it sufficiently to prevent it from launching aircraft, and quite another still to sink it. The carrier is the least vulnerable ship to destruction or sinking because of its large size in relation to the warheads constituting the threat, greater relative compartmentalisation, massive protection of propulsion and ordnance, and its structural strength necessary to handle the stress loads of catapulting and arresting 35 ton aircraft.[51]

He went on to assert that 'short of a direct nuclear hit, the carrier is the least sinkable ship ever built, and the larger its size the less its vulnerability'.[52] Given the fact that any carrier is going to be a relatively obvious target, the greater toughness of the large ship, as well as its greater fighting capacity, makes it the better platform. The supercarrier's Achilles heel is its vulnerability to underwater damage from mine or torpedo. If not sunk, it might well be put out of action and prevented from taking any part in further operations. A naval force's supreme asset would have become its greatest liability. This means that even supercarriers, formidable weapons systems though they undoubtedly are, must be deployed and used with a certain level of prudence.

Platforms

Only one navy has both the resources and expertise to operate this level of carrier capability in the foreseeable future. The Soviet Union seems to have evolved its large cruiser into an interesting new 60–70,000 ton carrier concept to operate both conventional aircraft and STOVL machines. The first two ships being built (the first of which is to be called *Tblisi*) retain many of the characteristics of their cruiser forbears, i.e. reliance on SSMs as their major ASUW capability and advanced SAMs for air defence. A ski-jump bow will allow the efficient operation of improved STOVL aircraft and a Yak-41 Harrier type aircraft is under development, but not without serious problems which may have enforced a reversion to experiments with more conventional fighters using catapults and/or ramps. Now a third larger carrier has been laid down. It is too early to say what will become of this programme but there seems little doubt that even if conventional fighters are carried the main roles of the ships will be AAW and ASW, and they will have only limited power projection compared with US ships.[53]

Although the American view seems to be that 40,000 tons is the minimum size of carrier for the operation of normal high-performance aircraft by means of catapults and arrester gear, other countries disagree. France plans to build new ships that are reported to have an intended full-load displacement somewhat below this figure. These vessels, nuclear powered like the new US and Soviet ships, will enter service around 1996 and 2001 to replace France's current pair of 22,000 ton carriers. The new design tries to achieve maximum utilizable deck area on the limited displacement. The new ships will not be able to launch and land simultaneously, due to the layout of the catapults, and they will not be capable of very intense operations. They will carry between 35 and 40 aircraft, a mix of the projected Rafale based ACM fighter bombers (the first ship might carry an interim air group of Super Etendards and F-18s), and helicopters. They will not be in the same class as US carriers but, in the many low-intensity operations in which France thinks of using its naval power, such an air group may well prove to be more than enough. As at present, demonstrative '*pré-stratégique*' use of nuclear weapons will be another important role for these ships.

For navies of similar, or lesser, means and facing relatively limited threats the smaller carrier still offers a most useful list of assets, as well as the prestige of having carriers in the fleet. A surprising number of navies operate small carriers and both the evolution of the helicopter and the development of the STOVL aircraft has enhanced this trend. Old aircraft of the 1950s and 1960s that could relatively easily be operated from modified light carriers

The changing shape of naval war

of World War II conception can be replaced at similar levels of capability by new STOVL aircraft, notably Harriers and Super Harriers. Smaller carriers come in a variety of shapes and sizes. They might be light carriers of an earlier age, soldiering on for as long as their refurbished engines or solid hulls can carry them. India's addition to her existing *Vikrant* of the *Viraat*, the 24,000 ton former British *Hermes*, fully modified for STOVL and helicopter operation, was a considerable vote of confidence in the longevity of even a ship built largely to low quality British wartime standards. Her 1986–7 refit gave her another ten years of active service. She possesses a modern combat information and C^3I capability: not for nothing was she used as flagship in the Falklands War rather than HMS *Invincible*, the first of Britain's three new 'aircraft carriers'. This was ironic as the latter ships had evolved as essentially air capable command/cruisers. Their aircraft-carrying capacity is about twenty, which is quite limited for ships of their size and expense, although attempts are being made to increase it. Ski-jumps have been added to increase the performance of embarked Sea Harriers. Although very useful vessels, especially with airborne early warning helicopters added to their air groups, the political factors that made a cruiser with its own integral SAMs, sonar, and command facilities grow into a carrier has produced a ship that is perhaps unnecessarily expensive compared to a vessel built purely for the operation of aircraft. It was lucky for the British that they still had a real carrier available to fight perhaps their last colonial war. The Invincibles are fully stretched in high-threat areas and would benefit from additional ASUW missile capability to allow their eight Sea Harriers to concentrate on combat air patrol and reconnaissance duties. Both Spain and Italy are going to enter the 1990s with new STOVL/helicopter carriers in the 10,000 ton (standard) class that will have more or less the air carrying potential of a British Invincible, but with fewer 'ship' features. Spain used the American 'sea control ship' design for the *Principe de Asturias* built in Ferrol. The slightly smaller *Guisseppe Garibaldi* was a wholly domestic product. It will be interesting to see if other current operators of old ex-British carriers of the same size, notably Argentina and Brazil, will find similar ships attractive or, more importantly, affordable; or indeed if other naval powers will show interest in small STOVL carriers. India has announced she will build her own new carriers in future but Japan still suffers from a political reluctance to procure such vessels.

It is possible to produce very basic carriers indeed by using fast merchant ship hulls. Such unsophisticated vessels lack much appeal to naval planners, who have many reasons (some rational, some less so) to build 'real' warships. It usually takes the pressure of actual operations to overcome these problems, and interestingly, Britain during and after the Falklands War made perhaps the most progress in modifying mercantile hulls for the operation of aircraft. Operated for some time afterwards was the helicopter carrier *Reliant*, using American Arapaho conversion gear and manned by the mercantile marine personnel of the Royal Fleet Auxiliary. She was eventually disposed of but has been replaced by RFA *Argus*, an extremely impressive ship with a large hangar, a protected flight deck, and high capacity. At a quarter the cost of an Invincible she is primarily intended as a training ship, although she would be operational in an emergency primarily as an ASW carrier in the Channel. Her ability to operate STOVL was limited mainly to save money and design complications but also perhaps partly for political reasons. A ski-jump fitted to *Argus*, although more expensive, would have raised some disturbing questions about relative capabilities given that the last Invincible class vessel, HMS *Ark Royal*, cost considerably more still.[54]

It would be idle to pretend that ships such as *Argus* offer a complete replacement for aircraft carriers of higher capability, especially in terms of sensors and command and control, but they do provide a means for a small power to provide an economical, if limited, air capability if it can afford no other, or for a larger naval power, perhaps even a superpower, to proliferate air capable platforms in a crisis, relying on the electronic equipment of escorting surface combatants to manage the battle. More capable carriers cannot be everywhere, indeed, as the US Navy's main instruments of power projection, they may well be sent on operations that have little to do with the essential sea control task.

Possibly the US Navy may itself move down the carrier scale and produce a ship of more moderate dimensions, as it came close to doing in the 1970s. The election of a president committed to military reform might mandate smaller carriers. These could be 45,000 ton mid-size carriers (CVV), 20,000 V/STOL support ships (VSS) or small 10,000 ton sea control ships (SCS). These ships would obviously supplement rather than replace the current leviathans.[55] As well as saving money their construction might reflect a conceptual judgement about roles and missions, and an emphasis on sea control in the Atlantic rather than power projection around the world. But for a whole range of reasons, not least the essential nature of US sea power, sea control has never struck too many chords among US naval planners. Given this professional ideology, coupled with the understandable self-interest of the naval airmen, it becomes much less surprising that the great bulks of the USS *Abraham Lincoln* and *George Washington* have been taking shape at

Newport News. Financial constraints on US Navy procurement plans could easily force concentration on these ships at the expense of small carriers rather than the abandonment of the supercarriers themselves. It is interesting that the Lehman study showed more interest in smaller ships as additions to US capabilities rather than as replacements.

In one sense, however, the US Navy currently deploys many lighter carriers. These are the various amphibious warfare vessels that are capable both of transporting US Marines ashore by helicopter, and also supporting them there with STOVL fighters and attack helicopters. This relatively little publicized fleet, any one member of which might transform the purely naval equation in many parts of the world, provides the alternative dimension of the USA's naval power projection ability. There are seven 11,000 ton LPHs (landing platform helicopters) and five newer 25,000 LHAs (landing helicopter assault). The LPHs in particular have proved useful beyond the amphibious role, being used as experimental 'sea control ships', i.e. limited capability ASW/AAW helicopter and STOVL aircraft carriers, and also as mother ships for MCM helicopters. This flexibility is being exploited in the new class of 26,000 ton LHDs (landing-ship helicopter docks), whose lead ship name, *Wasp*, a traditional carrier name, reflects a greater emphasis on flexibility as an aircraft platform. The new ship will be more easily convertible into an ASW carrier or, alternatively, an 'assault carrier' with short range STOVL ground-support aircraft. Some compromises have, however, had to be accepted between 'fixed wing' and amphibious helicopter assault capability, e.g. no ski-jump, in order to maximize deck space so that the maximum number of helicopters can be launched in an assault wave.

The LHDs, like the current LHAs, will have full docking capability, i.e. the capacity to flood themselves to act as sheltered docks for landing. Ever since World War II the US Navy has operated a huge fleet of such landing ship docks (LSDs) and landing platform docks (LPDs), the latter having a higher personnel carrying capacity than the former. The US Navy is complementing its new LHD programme by building ten new LSDs, each just over 11,000 tons in displacement. These will be capable, like the new LHDs, of operating the new LCAC (landing craft air cushion) hovercraft that are taking a more important role in amphibious landing operations. The hovercraft, with its ability to move with no difficulty from water to land, and to skim over the shore line, has obvious utility in the landing role and is finally being taken up by both superpowers for this duty, as well as by other states such as China.

Platforms

The USSR has nowhere near the long-range oceanic amphibious lift and power projection capability of the USA. She has a large force of small and medium landing ships in the less than 1,000 ton bracket, but these are only suitable for short-distance operations. They lend themselves to export and are in service with a number of friendly navies. Even her large landing ships are usually relatively simple tank-landing vessels, half the size of their US equivalents. The USSR only has two naval vessels capable of docking, the pair of Ivan Rogov class vessels. Given limited numbers, these ships have had to be made very versatile. They are capable of landing a battalion by hovercraft (or landing craft) and a small number of helicopters, and giving fire support with rockets and gunfire. The ship can beach if necessary to land tanks. Given the continued supremacy of the airborne forces in the Soviet Union's power projection capabilities, one cannot expect much extra development of the Soviet Navy's amphibious warfare capability.

Several other countries have LPD/LSD-type specialized landing vessels. Such ships are in fact probably essential if a country wishes to be able to land troops over beaches. Merchantmen can be most useful in amphibious operations as transports, as demonstrated in the Falklands War. Many Soviet roll-on roll-off cargo vessels seem to have been built with possible use as landing ships in mind. Yet exercises show that simple unmodified ferry-type vessels cannot be safely beached. Also, transfer to landing craft can be hazardous to men and ships alike. Military features, like beaching capability, must be built in at construction if the ship is to have operational functions beyond that of a mere transport.[56] This can be done with merchantmen built with rapid mobilization in mind, but the need for specialized command and control equipment mandates some specialized naval manned ships, as does the ability to deploy at short notice.

Unfortunately amphibious vessels tend to be relatively expensive. Even middle ranking states, like the United Kingdom which has deployed two LPDs since the 1960s, has to consider very carefully the question of replacement. She has decided in principle to replace these ships with rebuilds or new construction, and in addition to acquire at least one vessel for helicopter landings. Again the precise configuration and origin of the new ship has not been settled. She might be a converted merchantman, and British Aerospace and Sea Containers have produced an interesting 28,000 tonne 'aviation support ship' planned on the container ship *Contender Argent*, which would lend itself to the role. A 'through deck' ship (without ski-jump), she would be able to embark, transit, and deploy a Marine commando.[57] France plans to complete three new LPDs in the 1990s and Italy is constructing two new 5,000 ton San Georgio helicopter-operating landing ships. Interestingly, funds were obtained to build the second vessel, *San Marco*, by building for the Ministry of Civil Protection as a disaster relief ship. There are also smaller general-purpose landing ships on offer, e.g. the 2,000 ton, 76 mm gun armed vessels built by Brooke Marine of Lowestoft in 1979 or the 900 dwt LS100 design currently offered by Harland and Wolff of Belfast for Oman. Amphibious warfare vessels, with their large accommodation spaces, ability to operate smaller craft, and possession of helicopter flight decks that make excellent stowage spaces (and parade grounds), lend themselves also to use as training ships.

They also make good command and headquarters vessels. The United States has gone to the extent of building two specialized command ships especially for the control of amphibious operations. This pair of 20,000 ton vessels, based on LPHs, have a bewildering array of electronic command and control equipment, and are perhaps the most electronically sophisticated vessels afloat. They are beyond the reach (or the needs) of any other navy and tend to be used in peacetime as fleet flagships.

Before leaving surface warships, a word must be said about fleet support. If any navy wishes to have the capacity for sustained operations at sea it must be able to take supplies of fuel and other stores with, or out to, the operating forces and transfer them there to the combatant units. This means significant investment in relatively fast (20 knot) tankers and supply ships specially fitted to engage in replenishment at sea (RAS) operations. These can be supplemented by slower, less capable mercantile hulls in time of need. No navy can claim to be fully ocean-going if it does not have such ships, which can double as ASW helicopter platforms and aircraft maintenance facilities. There is a tendency to increase the combatant utility of these vessels thus integrating them with the whole force. This also implies a level of point defence armament such as Sea Wolf missiles fitted to the new British Fort Victoria class.

British Shipbuilders has recently unveiled a 'Multi-Role Naval Auxiliary' (MRNA) which is basically a replenishment ship but which has the potential for use as an amphibious warfare vessel with helicopters and landing craft, or disaster relief ship, oceanographic vessel, or MCM support ship. The ship has various extra survival and protection features compared to a normal mercantile hull and could be attractive to various navies. As its builders put it, given an MRNA's capacity to be configured in a number of various ways: 'a navy can procure a vessel tailored to its own particular needs. For some countries the MRNA can provide a navy with a new capability, for other navies it can provide an increase in capacity for undertaking various operations without the need for

Platforms

extra ships and crews.'[58] It will be interesting to see if such multi-role auxiliaries become more common.

China has recently built her first four fleet tankers of more or less conventional layout, the most important sign of her aspiration to longer-range surface operations. India needs to improve her capacities in this regard, a third fleet tanker is under construction but she, like the other two, will be chartered not owned. France, who has traditionally relied more on fixed bases, is also somewhat belatedly improving her capabilities in fleet supply ships, both by new construction and charter. Britain, who built up her supply vessel capabilities impressively in the 1960s to sustain her East of Suez peacekeeping role in a context of few shore bases, is allowing her supply fleet to run down, but is investing in limited numbers of new and expensive fleet replenishment ships, the first of which should be in commission in 1990. Soviet specialized capabilities look like remaining relatively limited, although the Soviet merchant fleet is always available to supply the navy when required. Inevitably, the United States has a supply of fleet auxiliaries second to none.

The final matter to be discussed is the question of surface ship vulnerability. Surface combatants are indeed inherently vulnerable to the widest range of enemy weapons. Thus they are not very good offensive platforms, except at the long ranges that carrier aircraft allow. However, they are not so vulnerable as to be swept from the seas by anti-surface missile threat. Hitting them at all is not so easy and little real damage may result even when a hit occurs. Modern surface ships have evolved from the traditionally non-armoured vessels of the past. Even the modern cruisers are the descendants of older destroyers rather than the older cruisers that used to carry a little armour against gunfire. The electronic sophistication of three-dimensional warfare and the need to maximize space for sensors, command equipment, and space-intensive missile systems have all emphasized ship designs that continue to optimize volume for a given displacement. Putting armour all over such spaces is a more difficult task than protecting the armament, mechanized vitals, and watertight integrity of 'old fashioned' Dreadnought era warships.

Much can be done to stop ships being unnecessarily vulnerable. Surface ship design is going increasingly to emphasize signature reduction, acoustics, radar, infra-red, and magnetic means. The physical shape of future navies is going to be increasingly conditioned by the need to make both electronic detection and identification difficult. Separate modular ventilation and fire-fighting systems can also greatly reduce vulnerability to fire and smoke threats. Aluminium, which if not quite so flammable as sometimes argued, is certainly much more vulnerable than steel to heat damage and which is also prone to splintering, will be used much less in warship construction. Plastics may, however, be used more, because of their 'stealth' characteristics. Less flammable materials can be used for wires, which themselves can be replaced by more modern forms of highway for electronic data distribution. Only the largest ships, however, notably carriers, can be armoured in anything approaching a traditional kind of way. In smaller vessels armouring would necessitate substantial extra increases in volume and power that cost would soon make prohibitive. Alternatively, the fighting power of the ships would be reduced, thus reducing their potential offensively and defensively.

It is likely, however, that the warships of the future will be given some small level of armour protection to limit excessive vulnerability to damage. Extra metal plating or other substances such as Kevlar can be added to especially sensitive areas, such as combat information centres. The increase in cost and weight is small, about 50 tons on a modern frigate. Even radar antennae can be given Kevlar covers for there is little point in protecting the ship if the means by which it can engage in above water warfare are disabled. The level of protection is not intended in any sense to protect the ship from the structural effects of direct hits, but it does limit collateral damage and the effects of near misses. It also protects against small-arms fire, important in a range of low-level situations.[59]

Submarines

Of course the major way in which a naval platform can protect itself from above water damage is to submerge deliberately beneath the waves. The twentieth century has seen the ever-increasing importance of the submarine, especially as nuclear propulsion has solved the fundamental problem of providing safe, reliable propulsion without the need for external air. Submarines have become the primary seaborne platforms for high-level anti-surface warfare. They are also extremely effective ASW platforms, being able to exploit the changing characteristics of the sea environment to optimize the performance of their sonar.

Submarines do, however, suffer from four very serious limitations. First, the kind of damage they can inflict with their primary torpedo or missile weapons is almost always fatal and catastrophic. This rules them out as weapons of much utility in operations at the lower levels of intensity. Submarines can be used as cover for more valuable forces in such situations and they can threaten unwelcome retaliation if the rules of limited conflict are broken; if they are unleashed then it usually marks a major escalation of the conflict.

Platforms

The *Belgrano* affair demonstrates the problem clearly.

Second, submarines are only relevant to the ASW and ASUW battle, they have very little AAW potential beyond a limited early warning role. Attempts have been made, and will be made in the future, to give submarines anti-air missiles and they might have some utility to a cornered, detected submarine, especially one being harried by helicopters. Submarines cannot, however, provide air defence for ships, either in terms of sensors or weapons, without giving away their position and thus losing their main advantage of stealth. Third, the submarine is unsuitable for roles where the presence of the unit, and the interest and the capability it implies, is the very point being made. Even when surfaced, submarines have the appearance of stealthy, silent killers which alienates rather than attracts. Only small submarines are good for port visits and there are obvious problems in the social side, although the ships' companies make valiant efforts to overcome them. Nuclear-powered submarines, whether armed with nuclear weapons or not, are rarely welcome visitors anywhere. Submarines do have excellent utility in providing a latent naval presence that can be revealed if required and denied if not, as skilfully demonstrated by the British Government in 1977 in relation to the Falklands, but they cannot substitute for surface warships in most elements of diplomacy.

Finally, and perhaps most importantly, submarines suffer from very severe command and control limitations. For precisely the reason that a submarine is so effective, its use of the sea to block electromagnetic radiations, it is extremely difficult to use the main communications medium, the same electromagnetic radiation. (Sound communication is only possible at the shortest of ranges.) Radio waves do not penetrate sea water to any extent, except at the lowest frequencies. Even with very low frequencies (VLF) the submarine has to put itself, or an antenna, close to the surface. Such low frequencies, however, have a very low traffic capacity and are only one way from the base complex and its vital transmitter to the submarine. Range is limited to about 1,500 miles. If a submarine wishes to have a dialogue it must raise its antenna above the surface and use higher-frequency radio. This, and the act of transmitting, makes it vulnerable to detection. The use of satellites and ultra-high frequencies (UHF) and super-high frequencies (SHF) can allow more private directional exchanges of information at very high speeds, and with much reduced chances of detection, but even this only works best on a routine basis as the boat must come shallow and extend an antenna. Expendable radio buoys can be used as alternatives to coming close to the surface. These can be released from submarines with pre-recorded signals or dropped by aircraft to allow UHF communication via sound signals. The latter, however, again makes the submarine vulnerable if the communication is to be two-way.[60]

Some of the problems of low frequency communication can be overcome by the use of extremely low frequencies (ELF) which increase sea penetration from 10–15 fathoms to about 100 fathoms. ELF allows staying deep, freer from detection by sonars or infra-red detection which can pick up buoys or wires. ELF, however, has a very low data rate, three characters in twenty minutes, thus making it merely an alarm to get a submarine to rise to receive more information. It also required considerable tracts of land, limiting its acceptability to the American public. An alternative means of communicating with submarines is to use blue-green lasers. These have the property of being able to penetrate sea water's 'optical windows'. They can be mounted in satellites and work has been going on in the USA on a Submarine Laser Communication Satellite System (SLCSAT) since 1978. This could allow much more specific and comprehensive communication from bases out to submarines, especially to ballistic missile firing boats. The cost, however, is considerable, some $2 billion at current prices, and the US Navy is proceeding relatively slowly with the programme. During the 1990s this method of communication could revolutionize submarine communications in terms of the degree of real time controllability of the submarine force, both SSN and SSBN. Given the uncertainty of the sea, however, and the availability of the right 'windows', SLCSAT would only supplement other systems: it would not be the sole means of submarine communication, VLF and ELF would remain. The US Navy has been accused of dragging its feet over SLCSAT and one reason may well be the instinctive service suspicion about the degree of control if would allow over its traditionally independent submariners. As one senior American official pointed out: 'the submariners of the USN must now accept that they were no longer "a law unto themselves"'.[61]

Poor communications mean that submarines operate best as 'lone wolves'. They receive basic orders and intelligence and then must be left to their own initiative, informed by general rules of engagement. This constrains matters at all levels, from the lowest, where it confirms the submarine's nature as a somewhat blunt instrument, to the very highest where ballistic missile submarines cannot easily or quickly be retargeted. It also makes close co-operation with surface units difficult, which is a basic problem given sea power still being largely about the use of the sea by surface platforms. Submarines can sink them but they have drawbacks as means of defending them. They are essentially weapons of sea denial rather than sea control.

For reasons explored at the outset of this chapter, it seems

Platforms

unlikely that submarine transports will drive naval warfare wholly into the depths for some time to come, if ever. Submarines will not, therefore, replace surface ships but they will remain potent naval weapons that fewer and fewer navies will feel able to do without.

Nuclear power, as mentioned above, has been the key to the submarine's quantum leap in capability. Not only does the nuclear reactor confer independence of the surface and make crew endurance rather than mechanical limitations the major constraint on a submarine's range, but it also provides massive power outputs that allow the most extensive and powerful sonar and command and control suites. Nuclear submarines are, therefore, vessels of some considerable size, about 3,500–6,000 tons for current American, British, and Soviet torpedo attack submarines (SSNs). Future boats will probably be in the 8,000–9,000 ton bracket. A need to carry bulky missiles causes elephantiasis to set in. Modern Soviet cruise missile submarines intended primarily for long-range anti-carrier warfare displace about 12,000 tons and the latest Soviet ballistic missile firing submarines are over twice as big. Double hulls make such 'boats' tougher to attack. The latest American ballistic missile firing submarines (SSBNs) are in the same large size bracket, being 17,000 ton vessels given the names previously reserved for battleships.

Smaller nuclear submarines have been built but they suffer from quite serious technical drawbacks. The French have limited their SSNs to a displacement of 2,500 tons, but the exceptionally small natural circulation reactor plant results in a lower maximum speed than that attained by most SSNs. This is a serious tactical disadvantage. Alternatively, the Soviets have built small, titanium-hulled Alfa class boats of about 3,000 tons with dense metal-cooled reactors and remarkable speed (45 knots) and diving capabilities. They were probably constructed as a temporary admission of defeat on the quietness front but construction seems now to have stopped. The boats were exceptionally noisy: this did not matter so much if the sole aim was to avoid destruction, but to do anything positive the Alfa would have to slow and/or come to a shallower depth to listen on sonar or to engage a shallower diving or surface opponent. The development of better Western weapons could also not long be delayed. Hence construction of the Alfa class boats ceased after six production units had been constructed and much consternation created amongst some more alarmist naval commentators. The Soviets seem to have paid considerable attention to developing sophisticated skins for their boats to minimize drag and hydrodynamic signature. It is clear, however, that the Russians have seen little alternative but to attack the problem of noise once more.

Helped by spies and Western firms willing to sell advanced constructional technologies they seem to be essentially copying the West, with relatively large boats with advanced propellers of non-cavitating design, propulsion machinery on 'rafts' to minimize sound transmissions, other machinery on flexible mountings, quiet reactor coolant pumps and anechoic coverings. The new Akula class submarine seems very much in the same size (8,000 tons dived) and technical bracket as a modern American boat, with the same kind of sound quietening techniques. Both the latest Sierra class boats, incremental developments of the previous Victors and the doomed liquid metal reactor fitted Mike (the latter apparently optimized for ASUW with torpedoes) were also quieter than their predecessors.[62]

Quietness is, of course, the crucial parameter in submarines, with an obvious impact on other performance characteristics. A submarine's various *quiet* speeds for optimum transit, search, area coverage, approach, and attack are much more important than an absolute maximum figure. The West still has the major advantages here, indeed the Americans now claim that their latest nuclear submarines are quieter even than conventional submarines. As well as 'rafting' machinery spaces, pump jet propulsion is used in new British SSNs. It is planned for the projected American Seawolf class of SSNs and can be used to diminish propellor noise still further.[63]

Mechanically, the pressurized water cooled reactor seems set to retain its supremacy for some time to come. Alternative types of reactor have severe safety drawbacks. Soviet style liquid metal reactors have been described by the head of the US Navy's Nuclear Propulsion Directorate as containing 'more radioactivity than all the reactor coolant in our whole navy'.[64] It has been suggested that the Soviets have operational magnetohydrodynamic waterjets on their latest submarines, but these reports have been discounted by authoritative sources elsewhere:[65] Even more futuristic propulsion concepts such as rippling skins seem equally unlikely to emerge from the Soviet production base. Electric drive of a more conventional kind may have more of a future however: it is inherently quiet and superconductivity techniques could greatly increase its efficiency.

Steam turbines, however, are likely to remain the major near to medium term form of transmitting the power of nuclear reactors to the driving shafts of new submarines, especially as new countries tend to develop SSNs. China, after considerable problems with development, has produced nuclear-powered submarines in limited quantities, both SSNs and SSBNs. This has no doubt been a factor in India's decision to lease nuclear-powered submarines from the Soviet Union: delivery of the first, a Charlie I class anti-ship missile firer took place at the beginning of 1988. At the same time, Canada

Platforms

was in negotiations with Britain and France for no less than ten to twelve nuclear-powered boats for service in the Arctic beneath the ice. Both Britain and France offered designs but the cost of the programme led to its eventual abandonment.[66] Other countries have expressed an interest in acquiring nuclear submarines. Brazil has started planning and spending money on a project to begin construction in the 1990s. Pakistan, spurred by India, is examining the possibility of conventional boats with small nuclear power plants to improve underwater performance. Such hybrid submarines may be a good compromise for nations of limited means.[67] There is also the problem that 'nuclear' is still a word with enormous political connotations, not least in Japan, a country sophisticated enough to build nuclear submarines if she wished.

Other than greater political acceptability, and despite their operational disadvantages, conventional submarines will always have one major advantage over nuclear boats: cost. They have other advantages too, standard of quietness better than all but the most advanced nuclear submarines. Diesel electric propulsion is standard for such boats with snorts to allow operation of diesels at periscope depth. Batteries provide limited range under water and propulsion with no tell-tale surface activity. Batteries of greater efficiency, and possibly, eventually, superconductivity techniques in motors and generators, will increase the underwater efficiency of such boats still further, although not in any way that would fundamentally change the nature of conventional operations, which emphasize low speeds and short dashes. Under development, however, are novel non-nuclear engines that could give longer ranges and higher speeds under water. Fuel cells, flywheels, Stirling external combustion engines using liquid oxygen, and the closed-cycle Krieslauf diesel are all under consideration. Sweden and West Germany, both countries with important political inhibitions about nuclear submarines, are leading the way in these developments, the former with Stirling engines, the latter with fuel cells.[68]

Constructing and manning submarines is probably more demanding than constructing or manning frigates, but the advantages of the submerged platform make submarines increasingly popular in the world's navies. West Germany, in particular, has done well with the export of completed boats and the technology to build them locally. The Soviet Union provides rather larger and less technically advanced boats to its friendly states, and China exports submarines also. A few countries operate large 2,000 tonners, usually obtained from the superpowers, but in the Dutch case such boats are constructed domestically, despite enormous cost over-runs and disastrous fires. Significant political pressure did not prevent The

Netherlands exporting advanced submarines to Taiwan which hopes to build them itself in the future. The other conventional submarines in the world tend to come in three size brackets, about 1,300–1,600 tons, about 700–900 tons, and about 400–500 tons. The last named are only suitable for use in coastal and restricted waters and are preferred in the northern European navies, but even here the tendency is to larger 900 ton boats as the usual compromise in requirement, capability and numbers. One detects a tendency for countries to be moving up a band in their conventional submarine procurement policies. Britain, previously satisfied with 1,600 tonner, is now producing a 2,400 tonner. Argentina is moving from 900 to 1,700 tons. One thousand now seems to be considered the minimum for open ocean operations, but Sweden prefers this bracket for its own impressive submarines specialized for use in the Baltic. Midget submarines are also making a comeback. Colombia, Pakistan, Taiwan, and Yugoslavia all have modern examples and the surveillance activities of the Soviet Union's varied fleet of small submersibles are notorious.

Submarines of all shapes and sizes provide excellent surveillance platforms, as well as means of attack. The latter have been enhanced by the advent of compact cruise missiles compatible with the tubes of existing submarines. These give submarines a potent new weapon against surface ships, and also against targets ashore. Towed sonar arrays can be used to give target acquisition potential. One must not underestimate, however, the power of the torpedo, which carries a much heavier warhead than any missile and attacks its target at its most vulnerable point, i.e. beneath the waterline. If submarines use their torpedo tubes to fire missiles, then missiles eat into already limited torpedo stocks. For this reason the USA prefers to fit extra missile tubes to its latest SSNs. The Soviets have had the choice made for them by the relatively large size of their missiles, although they have the extra potential given by the large torpedo tubes fitted to submarines to carry the large wake-homing torpedoes and the Soviet family of underwater ASW weapons. The Soviets also continue to use separate missile tubes for their large anti-ship and anti-shore cruise missiles that will be in service in submarines for many years to come. Certainly some cruise missile capability will be a vital part of the armament of the submarine in the future: reportedly the French have even given their latest SSBN a sub-surface launched Exocet capability as a self-defence weapon.[69]

Aircraft

Sea power and air power are indivisible. Seaborne platforms provide

Platforms

some of the most flexible, and least vulnerable ways of deploying air power, either in the form of aircraft or missiles. This has created bureaucratic problems, and accounts for the almost universally uneasy, if not downright bad, relationship of a nation's navy and air force. Such problems have often been compounded by the undoubted utility of land-based aircraft to operate over the sea in all three modes of naval warfare. Lines of demarcation have never been easy to draw and these disputes have sometimes had serious operational effects. The full potential of land-based air power over the sea has often not been fully exploited. Navies have been limited in their abilities to operate aircraft. The situation is usually happiest when the sea/land boundary is used as the basic organizational divide, i.e. aircraft designed to operate over or from the sea being operated by the navy: those over land by the air force. Historical accident, however, has often dictated otherwise. Certainly the aircraft specially intended for maritime operations are as much a part of the nation's 'navy' as any ship or submarine, an air force can also have a considerable relevance at sea as part of its general capability in attack or defence.[70]

Aircraft are inherently fast: they can cover great distances in short time spans, and strike rapidly and surprisingly. Aircraft can carry sensors to a height where they are not limited by anything other than their inherent technical characteristics, either to find targets for other platforms to address if communications are adequate, or to attack targets themselves. They can carry a wide range of weapons for the latter role: long-range missiles and bombs for ASUW, depth charges and lightweight torpedoes for ASW, missiles and guns for AAW. In the ASW battle they can even 'look' beneath the waves with the help of sonobuoys or magnetic anomaly detectors or, in the case of aircraft that can hover, active dipping sonar.

Their drawbacks are, like those of the submarine, the reverse side of the coin of their assets. Speed demands power and power demands fuel which takes up much of the available payload. Large aircraft can achieve substantial feats of endurance but this is measured in hours rather than days. Moreover, such aircraft tend to be inherently vulnerable. Very high endurance aircraft can be obtained by the use of light gases rather than aerodynamic lift. From the earliest days of naval aviation airships have lent themselves to operations over sea rather than land, and modern technology has overcome most of the problems which raised doubts over the safety and controllability of earlier examples. It is quite possible that airships will be used for certain naval tasks in the future, especially constabulary/patrol duties in economic zones. The airship's high endurance, coupled with its ability to stop and inspect, or even board, would be especially useful for constabulary duties, but it would also be very suitable as an airborne early warning platform with the radar in the airship envelope itself. Secretary Lehman was a great supporter of the LTA (lighter than air) programme, and in May 1987 an Anglo-American consortium was selected to build a series of 129 metre long helium filled, diesel powered 'patrol air ships' with a 30-day endurance (with refuelling). The programme became vulnerable with Lehman's departure, however, and without powerful support from an existing 'community' within the service it remains doubtful whether airships really will be appearing over US task forces after 1992, as originally scheduled.[71]

More conventionally, large fixed-wing aircraft will certainly remain a valuable asset, both for surveillance and warning duties, against three-dimensional threats, and for attack against surface and underwater targets. Although the combination of a relatively simple airframe with a relatively unsophisticated radar is adequate for constabulary surveillance duties, capability at a higher level of maritime warfare demands much more sophistication and expense. The specialized maritime patrol aircraft has evolved as a highly expensive combination of electronically co-ordinated and analysed radar, sonar and electromagnetic sensors, combined with the ability to deliver depth charges, lightweight homing torpedoes and anti-ship missiles. These aircraft are currently best represented by the American P-3 Orion and British Nimrod, both based on airliner airframes, the Electra and the Comet respectively. The United States evaluated three contenders for its new 'long-range air ASW-capable aircraft' (LRAACA): Lockheed, Boeing, and McDonnell Douglas each offered different concepts. Boeing used turbofan engines in a derivative of the 757: these power plants now offer comparable fuel economy and range to the improved turboprops used by the developed Lockheed P-3 that was eventually successful in the competition. (In fact the Nimrod solved the problem of jet propulsion for maritime aircraft many years ago by its facility of cutting engines in and out depending on the speed required.) The McDonnell Douglas design utilized the new unducted fan concept that combines the speed of the turbofan with economy approaching that of the turboprop. The US Navy insisted that its future LRAACA should be able to stay on task for at least four hours, with a minimum radius of action of 1600 nautical miles and played safe with an advanced turboprop for its new aircraft that will be in production until 2001.[72]

Whatever the layout, such ASW aircraft are complex and expensive items. The simpler anti-surface task can best be addressed by long-range bombers equipped with less extensive sensor suites. The characteristics required are those of any long-range strategic bomber, and the Soviets have made this form of long-range maritime striking power very much their own. The USAF is, however, also demonstrating greater interest in a maritime role for its strategic bombers as penetrating the Soviet Union with them becomes more problematical. All such large aircraft, however, given their inherent vulnerability, must rely on stand-off missiles of various ranges to allow them to strike a well-defended target. One future concept that has been suggested to exploit these weapons is the long-range Missileer, a large 747-type aircraft crammed with offensive missiles of various ranges, and with laser weapons for self-defence.

Smaller aircraft facilitate greater penetration as well as being much more suitable for anti-air duties. Given the relatively low payload of such machines, range is limited unless substantial tanker support is provided. Sea basing is probably a better answer to this problem, although modern and future combat aircraft of the highest available capability will require large and expensive aircraft carriers of conventional type. The United States will remain the only navy with the ability to deploy such vessels and is developing an advanced, stealthy attack aircraft, the A-12, to take over the all-weather attack role in the mid-1990s. This may well become a common aircraft with the USAF, whose Advanced Technology Fighter may replace the F-14. For the other carrier aircraft roles – AEW, ASW, and electronic warfare – a common advanced tactical support aircraft (ATSA) is due for development in the 1990s.

All the above are conventional take-off and landing (CTOL) concepts. If reduced performance is acceptable, however, then more unconventional layouts can be used to facilitate sea basing. Vectored thrust jets have already allowed the deployment of relatively low-performance STOVL jet fighters which, when combined with high technology missiles, have demonstrated an ability to defeat even a relatively well-equipped Third World air force. The Harriers that achieved this feat remain the only effective STOVL aircraft of fighter-type performance. Produced in the USA as a clear weather fighter bomber, the concept has been developed there into the higher performance AV-8B. Britain has produced the radar-fitted Sea Harrier with lower flight performance but more all round avionic capability: the Sea Harrier is currently in process of upgrading to FRS-2 standard with better radar and long-range radar homing AMRAAM missiles. Sea Harriers are in service with the Royal Navy and India; AV-8s in various forms are in service with Spain and the US Marine Corps. It is not clear which branch of the family Italy will purchase. Japan is also interested in a Harrier-type purchase, as is Brazil, and the widening market is encouraging the joint manufacturers, British Aerospace and McDonnell Douglas, to consider a Multi-Mission Harrier II which will combine an uprated engine with radar. If decided upon it will enter service in the early 1990s.[73]

Such aircraft put usable air power within the means of a widening range of middle-level states. For the superpowers, STOVL technology in various forms is also probably going to be exploited more. The British developed the 'ski-jump' take-off aid, which allows vectored thrust aircraft to take off with higher payloads of fuel and/or weapons. It is now to be fitted to the new Soviet nuclear-powered aircraft carrying supercruiser *Tblisi*, currently under construction. The crude VTOL Forger aircraft used from the Kiev class cruisers must be hard to use with ski-jump, but it seems that the new Yak-41 may overcome this problem if it conquers its own difficulties. The Soviets have persevered with experiments to convert conventional take-off and landing (CTOL) fighters to carrier operations (using both catapults and arrester gear and ski-jumps). The fate of these experiments will be governed by economic and political factors as much as technical ones.

On the American side, US Marine Corps interest has allowed the diversion of some of America's massive resources into limited STOVL development. Aviation journalist Bill Sweetman has, however, summed up the problem succinctly and not too misleadingly as regards a more extensive US espousal of STOVL technology:

> The world of naval aviation is divided into the U.S. Navy with its mighty fleet of supercarriers and the rest of the world with its handful of ships per country. The rest-of-the-world market is too small and scattered to sponsor extensive development programmes; the U.S. Navy is so terrified by the prospect of armchair admirals promoting STOVL as a substitute for the supercarrier that it finds it difficult to consider it as anything but a threat.[74]

The problem is not only bureaucratic. Supersonic STOVL designs have been considered for many years but there are real technical problems giving vectored thrust aircraft afterburning engines of sufficient power. In addition, large and heavier aircraft require more complex control systems. Alternatives have been offered to vectored thrust, but none is yet practical. The most likely outcome seems to be an afterburning (or 'plenum chamber burning') development of the Harrier concept, followed in the next century by the exploitation of second-generation technologies like the 'tandem fan' which splits the airflow from a conventional augmented turbofan. Vought have produced a naval fighter design, the TF-120, based on this principle. As advanced STOVL types approach the performance levels of CTOL aircraft the limits on the latter's development may well help

STOVL types to 'catch up'. This will probably make deployment of significant ship-borne aviation an available option to all the major naval powers of the twenty-first century.[75]

Propeller-driven STOVL and VTOL technologies with considerable potential are already becoming available. The most important is the American Bell-Boeing V-22 Osprey tilt rotor aircraft which was to be delivered to the US Marine Corps for amphibious landings and to the US Navy for search and rescue, special warfare, and fleet logistic support. It could have also been developed to provide useful ASW and airborne early warning aircraft to operate from a wide range of platforms. Very sadlly the whole project has been cancelled to save money, it is to be hoped only temporarily, as such aircraft offer greater speed, range, fuel efficiency, reliability, and maintainability as compared to helicopters. The V-22's twin turboshaft engines allow it to cruise at 275 knots, and as a troop transport it has a range of 500 nautical miles. Although initial procurement costs are high, operating costs should be lower. In the longer term, jet-powered VTOL multi-mission aircraft have been suggested. Again these will make the development of a militarily highly capable STOVL air group a practical possibility.[76]

Some see the future belonging to STOVL or VTOL aircraft operating from all ships. A Skyhook technique has been developed in Britain to allow Harrier operations from destroyer type vessels. Skyhook is a 'clever crane capable of capturing a Harrier in the hover and placing it in a prescribed position at its base'.[77] The technique is technologically both practical and original, but it seems unlikely that destroyers will soon acquire STOVL aircraft as part of their armament. Such fighters need considerable logistic and manpower support. They must also be carried in numbers if sustained operations are required: and the most cost-effective solution, both logistically and tactically, is to group them in one ship with a large hangar and, therefore, a large flat deck for a roof. It would seem sensible, therefore, to develop specialized carriers for such aircraft rather than trying to equip the entire fleet with them. Such carriers, however, might be of novel design. Conversions of mercantile hulls have been discussed above. Indeed Skyhook technology might allow small container ships to operate as forward operating bases or as escort carriers in their own right. Alternatively Skyhook might make small new construction 'air capable destroyers' in the 5,000–6,000 ton class more practical. Effectively small escort carriers with short 'through decks' and 'ski-jumps', they would allow countries with political problems like Japan to obtain the benefits of STOVL capabilities. Their complement might be five Sea Harriers plus a large helicopter. Both the coupling of air capabilities with relatively

The changing shape of naval war

small displacement and the imaginative adoption of novel foreign technology are things which fit Japanese culture and experience extremely well.[78]

This is not to say that low-performance VTOL aircraft will not allow virtually any platform to be used as a forward operating base. Already the helicopter has achieved a remarkable degree of ubiquity at sea. Conventional rotary-winged machines provide VTOL platforms of sufficient range and performance for ASW surveillance and attack, limited ASUW surveillance and attack, and a wide range of utility and logistical duties. It is now possible to operate even large helicopters on ships of frigate size and above, and any sacrifice of such abilities is a considerable degradation in the ship's capability. The new importance of active sonar will emphasize the helicopter's usefulness as a sonar platform still further.[79] Helicopters will continue to be a vital part of the naval scene, acquiring new roles as target designators for laser-guided ship-borne artillery. They will also continue to be used for amphibious operations, although the tilt-wing aircraft and the hovercraft may cause a long-term decline in the uniqueness of the helicopter's current ability to carry troops and stores inland from offshore platforms.

Helicopters can also provide airborne early warning facilities, although their vibration and endurance characteristics make them less than optimal for this work. They cannot carry the sophisticated electronic computers necessary to back up a modern AEW radar. Probably, Osprey-type aircraft, with more load carrying capacity, will have greater potential in this role, providing a more capable AEW to the STOVL carrier equipped navy. Certainly, as the Falklands showed, some AEW is vital to the viability of any use of the sea against the threat of air attack, especially by small, fast, low-flying aircraft. Conventional carriers, like the US Navy's, can carry fixed wing AEW aircraft of maximum capability, but again this is one resource that is perhaps uniquely relevant. Otherwise, land-based, high-endurance AEW aircraft can be provided, if available, and can be useful to any maritime commander: they could be optimized for naval purposes, being fitted with weapons for ASUW strike.

Here, however, we return to the question of the allocation of 'air power' assets. Aircraft can be extremely effective over the sea, but only if properly allocated, committed, and trained. When any of these three latter attributes has been missing the result has often been disastrous. Given the ever-growing expense of aircraft, especially those with the greatest potential and utility (e.g. long-range AEW machines), there must be clear agreement between the various operators on roles and rigorous training in them. Even so a limited

Platforms

insurance policy (e.g. the British solution of AEW helicopters) is advisable to provide adequate assurance of some AEW support in the case of unexpected events, which tend to be inherent in the maritime environment.

Unmanned platforms

The 'unmanned platform' term covers remotely controlled vehicles operating above, below, or on the water's surface. Remotely piloted aircraft ('drones') have not been exploited much for maritime operations, due perhaps to the fiasco of the American DASH (Drone Anti-Submarine Helicopter) system in the 1960s. Manned helicopters proved a more practical answer to the particular problem of long-range ASW weapon delivery. Reconnaissance drones were also experimented with in the Vietnam War but recovery at sea proved difficult. More recently, however, after positive experience with Israeli-built drones as spotters for naval gunfire support, the US Navy has begun development of drones for a number of naval applications.[80]

The most important application of such remotely-piloted vehicles (RPVs) seems to be the form of obtaining over-the-horizon targeting data for the long-range weapons systems now becoming available. Such weapons also demand more lofted sensors to deal with incoming threats. Although the technology for such devices is now increasingly available, recovery is still a crucial limiting factor. Recapturing an RPV at sea is a tricky business even in good weather conditions. As RPV weights increase, the possibility of ship damage goes up in proportion. Ditching RPVs in the sea usually causes an undue level of damage to the vehicle. Unless an RPV is to be considered expendable (a possibility for certain applications) then a solution to the problem must be sought. One is to use tethered systems, such as lighter-than-air 'aerostats' already used by the US Coast Guard for surveillance. More mobile remotely-piloted lighter-than-air surveillance vehicles are also conceivable, with an air endurance measured in weeks between refuelling and maintenance periods carried out while hovering over surface ships.[81]

It seems likely that remotely-piloted aircraft of various kinds will overcome their technical problems in the next few years. In navies with significant air lobbies (either fixed or rotary winged) there might well be opposition to their deployment, but the USN is already deploying the Pioneer heavier than air RPV system for fire control in its battleships and some of its destroyers. The ever increasing need for relatively long-range surveillance and target acquisition and designation should force the increasing adoption of such vehicles at sea.

On and beneath the sea, MCM requirements are already causing various nations to develop and deploy remotely piloted surface and underwater vehicles of various types. More general deployment of robot submarine platforms also seems likely. One British proposal, the Seicon Patrolling Underwater Robot (SPUR) concept, foresees a vehicle with considerable autonomy guided by artificial intelligence. It would be used both for information gathering and 'suicide' attacks. The US Navy has concepts for 'small mobile sensor platforms' deployed from torpedo tubes for target detection and assessment. An American 'unmanned underwater vehicle' (UUV) prototype 40 feet long and 4 feet in diameter is being developed, which is too big for submarine deployment. Norman Friedman has speculated that this might be carried by a surface ship or helicopter and make a suitable bi-static receiver to give a 'three dimensional sonar picture. Alternatively it might be used to service or to plant successors to the SOSUS fixed sea bottom arrays perhaps in enemy waters'.[82] Successor UUV systems might also be used as transmitters for multi-static sonar: this would not risk the parent vessel, especially if it were a submarine. The usual problem of submarine communications creates difficulties for remote interaction with these subsurface platforms, but these are not insurmountable. Fibre optical cables would be useful over relatively short distances while sound signals (perhaps via air-dropped UHF buoys) and/or VLF radio might give longer range communication. This would be tricky, however, in the under-ice operations often mentioned in connection with these vehicles. Much would be expected from their on-board computers and data-banks.

Space platforms

Increasingly, space platforms will interact with naval operations to provide an extra dimension to the already complex sea power situation. As the last US Chief of Naval Operations put it: 'Sea control means space control.'[83] Orbiting satellites provide global UHF and SHF communications coverage that is less susceptible than normal radio to detection, jamming, and atmospherics. These communications are absolutely essential for the execution of modern multi-platform naval operations and their further improvement in terms of resistance to detection and jamming remains a high priority. The use of laser beams for communication will help solve these problems as well as allowing penetration into the sea to submerged submarines. Navigational satellites, also, will give increasingly precise fixes to naval forces. Such satellites can also be used to provide guidance data to missile systems, both cruise and ballistic.

Platforms

Surveillance satellites can already help solve navigational, and even operational, problems by giving meteorological information, but the most important way in which these devices will impact on naval operations is by watching for, and keeping track of, enemy naval forces themselves. There are two ways in which this can be done, either passively watching for electromagnetic radiations (e.g. radars) from ships (Electronic Ocean Surveillance Satellites or Eorsats), or by the use of active radar to detect seaborne objects (Radar Ocean Surveillance Satellites or Rorsats).

The US Navy has access to various intelligence-gathering satellites operated by the Central Intelligence Agency and the US Air Force. These, however, tend to concentrate on fixed points on land rather than moving targets at sea, and the US Navy, therefore, began in the early 1970s to experiment with its own surveillance systems, at first attached to USAF platforms. These culminated in the deployment of the USN's own Naval Ocean Surveillance Satellite Eorsat system. This is based on a set of White Cloud satellite clusters integrated with five major receiving stations and into a system of ground processing known as a whole as Classic Wizard. The satellites pick up not only radio and radar emissions, but also millimetre wave and infra-red radiations. The latter can be used to give images of the ocean's surface and indications of what is happening underwater. The elements of the system interact with other forms of electronic intelligence gathering to produce in the Naval Ocean Surveillance Information Centre a comprehensive picture of what is the situation at sea, or more accurately what it was recently. This lack of real time data is a major defect in the system. Ships can move away in any direction from the position in which they were originally detected. Other problems are the arbitrariness of some of the signal processing. Targets picked up on the edge of a cluster's area of interest are sometimes placed at the wrong point in the final output. The information received from the sensors is insufficient to detect submarines with any degree of reliability. All the system can currently achieve is the provision of basic data from which a part of the strategic picture can be made up for transmission as general guidance to naval units. It can be of great help to the naval commander, but the danger is that the magical information from the Eorsat satellite is sometimes believed rather than the visual data of the man on the spot!

In order to provide tactically useful data more promptly, in 1977 the US Navy began developing a system called Clipper Bow to give seaborne commanders 'all-weather/day-night active global coverage for surveillance and targeting of hostile ships in the critical area around the task force'.[84] The aim was to produce a high-resolution active radar satellite system, but the project ran into trouble in Congress, partly due to the technical limitations, notably an inability to deal with aircraft as well as ships. A revised requirement for an Integrated Tactical Surveillance System that can integrate satellite data with other sources was issued in 1981. This requires satellites that can deal with both ship and high-speed air targets. Such Rorsats should be in orbit by the 1990s. High resolution should be encouraged by the use of synthetic aperture radar (SAR) techniques, as was experimented with by NASA's SEASAT-A experimental ocean survey satellite in 1978. If the Strategic Defence Initiative leads to the orbiting of a world-wide surveillance system the US Navy might take advantage of such a facility.[85]

The Soviet Union has already orbited radar ocean surveillance satellites, although not without its own problems. Soviet Rorsats use nuclear power plants, and this caused particular embarrassment when Cosmos 954 came down over Canada in 1978 and Cosmos 1402 broke up in late 1982 and crashed into the Indian Ocean early the following year. Given the Soviets' requirements to find and target large Western ship targets and their doctrine of mass stand-off missile attack, they were less concerned than the Americans with providing high resolution for their Rorsats. The Soviets also deploy passive electronic intelligence satellites, sometimes in pairs with Rorsats. Satellite surveillance is especially useful to the USSR to provide targeting data for the latest Soviet high-speed, long-range anti-ship missiles.[86]

As well as providing increasingly comprehensive intelligence information, satellites may be used to participate more directly in maritime operations. The development for ABM purposes of directed energy weapons in satellites, or orbiting mirrors to direct ground emitted laser energy, opens up possibilities for the use of such systems against naval targets, if not to damage platforms themselves, then at least to impair their sensors. At a slightly exotic level, the increased dependence on satellites for naval communications, navigation, and surveillance add a new edge to the debate on the development of anti-satellite weapons (ASATS). A campaign against satellites could have a serious impact on the battle at sea. The naval commanders of both superpowers might feel they had an interest in maintaining a capability to knock out such key systems in the direction of the other side's naval forces. Alternatively, an ASAT ban might allow both sides to exploit space even more and in greater security in order to increase the effectiveness of their naval forces.

Other nations, such as the United Kingdom, have deployed their own naval communications satellites, and the diffusion of space

Platforms

technology promises eventually to make some ocean surveillance capability available to navies other than those of the superpowers. The increased transparency of the future is bound to diminish the surprise and concealment traditionally associated with naval operations. The nature of the maritime environment, however, with all its parametric variations, makes it unlikely that even the superpowers will always be aware of everything that is occurring over, on, and beneath the vast spaces of the oceans of the world. And even if they are, many others will not be.

Interconnections among the platforms

As Norman Friedman put in in 1986, 'it may be more accurate to characterise the capabilities of the Navy not only in the traditional terms of platforms, but also in the interconnections among the platforms'.[87] The growing complexity of twentieth-century naval warfare has progressively emphasized the synergy of the various naval force elements. Ever since World War II, surface warships have been able to use aircraft under their direction and control as an integral part of their 'armament'. Now the electronic revolution in the technology of information and distribution allows still more intimate integration. As essential feature of the automated AIO arrangements, described in Chapter 5, is their ability to converse with each other using data links. This allows one ship to target its weapons on the information of another, perhaps some distance away: thus 'a fleet can become a unified weapon system in which different capabilities are distributed among its units'.[88]

This is not quite so straightforward as it sounds. Not all ships are equipped with tactical data systems, or equipped with link equipment. The interface between different navies is sometimes difficult to achieve. Jamming is a constant threat, at all frequencies. Nevertheless, the various platforms of the most advanced navies will increasingly be capable of fighting in a remarkably integrated way. This might well make up for shortfalls in the provision of the most expensive types of naval assets, e.g. Aegis-type systems could be used to fire the missiles of a much less well-equipped ship. Even more so will it be impossible to assess the naval power of a state by simply assessing the individual members of its naval forces. The total force posture, and its ability to act together as integrated units on the basis of the widest possible foundation of available sensor information, will be the more reliable guide.

Part IV
The evolving environment

Chapter 7

The international and legal context

In Chapters 7 and 8 we examine the evolving environment in which sea power will operate over the next few decades. Chapter 7 deals with the political interactions of states which at sea increasingly involve a legal framework. Inevitably economic questions cannot be excluded. One does not have to be a Marxist to accept that economic factors are the foundation of national power and policy. The more sheer resources possessed by a state the more foreign policy instruments, both military and non-military, there are available. This chapter must, therefore, concern itself with a broad assessment of the economic foundations and prospects of the nations of the world and how they will affect the international politics that will be its major subject matter. Chapter 8 will address itself more narrowly to the economic factors that stimulate nations to have interests in the oceans, interests which may be worth defending militarily, as well as the factors which constrain their maritime military postures.

The world international system has been dominated since 1945 by two superpowers, the USA and the USSR. The former has constantly outperformed the latter in all areas except, arguably, the military. The United States looks like remaining the possessor of the world's largest gross national product: by the mid-1980s it stood at $4,000 billion, about twice that of the USSR. Japan, however, has caught up, and, depending on one's dollar assessment of the Soviet GNP, may well have overtaken the lesser superpower. Japan's high growth rate should confirm her as the world's number two economy in the 1990s, although some analysts estimate that China may have overtaken her by about 2010.[1] Generally in industrial Asia, economic growth rates are high; European countries continue to perform relatively well.

Both current superpowers seem to be in serious trouble economically. The United States is facing high balance of payments and trade deficits. From the world's leading creditor she has become the world's leading debtor. Curing the USSR's problems is the

The evolving environment

essence of Gorbachev's *Perestroika* policies. Relatively, therefore, other countries look like having a greater share of the world's production and income: they will, therefore, become more important. Superpower hegemony will be mitigated and political multipolarity enhanced.

Another effect of this process is that on the Soviet–US confrontation itself, especially in Europe. As Jonathan Dean has pointed out: 'The peak of the East–West confrontation in Europe has probably passed. . . . There is a possibility over the next twenty years of the gradual decline or attrition of the confrontation under the combined impact of arms control, political measures and budgetary shortages.' Ambassador Dean goes on to conclude that

> European confrontation and the rival military alliances that organise it will be with us for many years to come. But there is no requirement for the confrontation to be at the present level and size, cost and risk. To the contrary, now that the high point of the confrontation has been passed, the task is to move deliberately but also cautiously towards dismantling it to the extent possible.[2]

Dean argues that the descent from this watershed must be managed carefully, a theme taken up by Stan Windass and his colleagues in their work on 'Common Security in Europe'. They argue that a safe descent requires two 'skis', a strong defence policy and an imaginative co-operative security policy, the latter being negotiations between the adversary partners on arms control, confidence-building measures, and crisis control.[3] This process is liable to place an increased emphasis on the relative importance of naval forces; sea-based nuclear systems will probably remain the least vulnerable and most stabilizing strategic delivery vehicles. They will probably tend to be emphasized in post-treaty balances. Similarly, as deterrent strategies move towards increasing reliance on their conventional component, naval forces become more important to sustain the potential operations of theatre conventional forces (see Part V).

The increasing emphasis on co-operative security will, however, pose something of a challenge to navies. Although naval arms control has a long and relatively distinguished history, having played a significant part in 'facilitating'[4] the transfer of naval supremacy from Britain to the United States in the inter-war period, the West has been markedly reluctant to accept agreed constraints on naval forces, deployments, or activities since 1945. Only SSBN forces have been affected by SALT Agreements and they will be included in a new START Treaty when it eventually materializes.[5] Conventional naval forces and operations have, however, remained largely unconstrained. Indeed, the West went out of its way to exempt 'independent' operations from the confidence-building measures (notification and observation of exercises and other significant military activities) discussed at the CDE Talks in Stockholm in 1985–6, and has refused to discuss naval forces in the new negotiations on conventional forces in Europe. The only naval operations included in the CDE Agreement were amphibious landings. If amphibious exercises include 3,000 ground troops or over they have to be notified forty-two days in advance: if they include more than 5,000 troops, observers have to be invited.[6]

John Borawski, Director of the Political Committee of the North Atlantic Assembly and a leading expert on CDE and CBMs in general, sums up the Western view on independent naval activities thus:

> Not all naval activities surrounding Europe affect European security directly (the Falklands War or U.S. operations in the Mediterranean directed towards Middle Eastern states) and . . . in any event, . . . defining naval 'manoeuvres' would be difficult because 'manoeuvring' is a routine aspect of naval operations. The threat to Europe, it was further argued, arose primarily from ground forces and thus they should receive primary attention. Another concern was that the Soviet notification measures would set a precedent for constraint proposals, such as for a nuclear free Mediterranean or geographic bounds on aircraft carrier deployments – a presumed Soviet objective in pursuing this line of proposals.[7]

The Soviets recognized Western reluctance and at the beginning of 1986 announced the 'deferral' of consideration of independent naval operations at the CDE. The USSR, however, which seems to perceive it has more to gain from such proposals than the West, will not abandon the demand for naval co-operative security measures. Indeed, from 1986, Soviet spokesmen, including Gorbachev himself, produced an impressive 'menu' of such proposals. These included the mutual withdrawal of Soviet and American naval forces from the Mediterranean, the placing of limits on the movements of ships equipped to strike coastal areas with nuclear weapons, and a ban on ASW measures in ballistic missile submarine 'sanctuaries'. All these measures ran counter to Western naval strategic thought and also raised considerable problems of verification for the West. Nevertheless, the Soviet 'menu' also contained serious proposals which stand more chance of forming the basis for agreement. Soviet spokesmen have called for agreements not to conduct naval exercises in international straits or other agreed areas, limits on the numbers of naval exercises held annually, and the advance notification of exercises and the invitation of observers. Various threshold figures

The evolving environment

have been suggested for notifiable or constrained activities. At Stockholm the Soviets suggested 30 ships and 100 aircraft as the size of exercise beyond which notification would be mandatory. In November 1986 Mr Lygachev, then number two in the Politburo, put the level at 25,000 men (a large carrier battle group with two carriers might contain about 15,000). It was suggested that an exercise greater than this figure should not be held 'more than once or twice every one or two years'.[8]

Such proposals as these might well form the basis of a regime of naval confidence and security building measures that could evolve over the next decade. Indeed, unless the West shows some (albeit necessarily limited) recognition of Soviet maritime concerns, the whole conventional arms control process might make but slow progress. There are useful precedents for purely naval co-operative security measures. Both the United States and the United Kingdom have signed bilateral 'Incidents at Sea' Agreements with the Soviet Union, which regulate operations when the ships of those states are in close proximity so as to avoid the risk of either collision or accidental combat. Weapons are not to be trained at each other, nor objects launched in the other's direction, nor searchlights shone to dazzle those on the opposing bridge. In order to reduce the risks of collision both treaties include agreements not to 'conduct manoeuvres through areas of heavy traffic where internationally recognised traffic separation schemes are in effect'.[9] This is an interesting precedent for wider CSBM (confidence and security building measures) application.

As East and West become further entwined in this web of common security their confrontation may well become gradually less significant as the primary focus of both their naval rationales and their naval deployments. The focus might well shift to the utility of naval forces in sustaining national interests on a more global scale.

It seems unlikely in a world of sovereign states that global conflict will become any less endemic in future decades. As Ken Booth has put it: 'The periodic explosion of international conflict is one of the few things that can be taken for granted in an unpredictable world.'[10] Sir James Cable would put things a little more strongly: 'The world is not becoming a more peaceful, orderly or law abiding place. When comparing conditions from 1920 to 1939 with those after 1945, the growth of lawless violence is striking and progressive. More than ever before governments today are prepared to use force not only against their own nationals, but also against foreigners. And these are not the governments which started most of the trouble in the past.'[11]

Many conflicts, especially those not directly involving *both*

The international and legal context

superpowers, do not have much perceived potential to raise the dreaded nuclear stakes. Intervention operations by the superpowers to restore a favourable situation within their acknowledged spheres of influence seems still to possess a legitimacy which shows little sign of going away. This takes a maritime form in the American Caribbean sphere of influence, where the Grenada landings of 1983 provide a classic near contemporary example. This American sphere is less sharply delineated than the Soviet land empire and the Americans show somewhat greater restraint in 'sending in the Marines' than they did before 1939, a restraint from which contemporary Nicaragua benefits. Nevertheless, it would not be surprising to see future Grenadas, given the right circumstances of chaos and power vacuum ashore. When things are not quite so chaotic, naval forces may still exert pressure, even if direct intervention may be more difficult both politically and operationally.

Sir James Cable is right to place the degree of organization of modern governments above the sheer military capability provided by modern weapons as the major factor making applications of great power naval force more difficult.

> Few nations can claim an increase in political wisdom during this century, but many governments now have more control over people and resources. By the narrow and sinister test of their ability to sustain organised combat, the governments of older and more civilised countries have lost much of their earlier advantage. In terms of political and military discipline, they are closer to a rough equality with lesser developed nations.[12]

This higher degree of organization goes hand in hand with a greater degree of regard for national sovereignty. In a world where at least lip service is paid to the UN Charter, the political legitimacy at governmental and popular levels of one country using blatant force or threats of force against another has significantly diminished. This does not mean, however, that states will always deny themselves the option to intervene by sea. Ken Booth sums up the balance of factors well:

> International politics is always changing, but more kaleidoscopically than in a linear direction. 'Intervention is dead' proclaimed the war in Vietnam. 'Long live intervention' proclaimed some Africans in the 1970s despite the rhetoric of the OAU. Colonialism had gone, but a form of tacit neo-colonialism seems to have arrived. The weak will always look for protectors; and the strong will always look for gain. Interventions will therefore continue to take place. But if they will not be as frequent

as in the historic 'age of imperialism', because of changing cost-benefit calculations, they will certainly be more frequent than was generally assumed during the anti-intervention mood which characterised Western thinking in the years following the American tragedy in Vietnam. The continuing 'agony of the small nations' will provide frequent opportunities in the future for powerful states to intervene for ambitious purposes, to intervene as a result of feelings of insecurity, to intervene in response to requests, to intervene to pre-empt another's intervention and to intervene to maintain the principle of non-intervention.[13]

The United States has been the major recent practitioner of maritime intervention and may well remain so.[14] Her network of world-wide commitments and interests, and the simple fact that these are truly *overseas* commitments and interests will mandate a forward power projection role. Even though American naval power may begin to be constrained by the twin factors of deficit reduction and a slightly less forward strategic posture in the post-Reagan years, the fact of America's traditional superpower position and the perception that the Western buck stops in the White House, will continue to pull her into various uses of the US Navy and Marine Corps when they are considered appropriate. Such uses could escalate into large-scale military operations in which shipping would be vital for logistic support.

The Soviet Navy, even at the height of the Gorshkhov era, was always less ready than its superpower rival actually to fire its guns or missiles in anger. It tended to be used more to neutralize American options than act as the main instrument of Soviet intervention, although this is not to underestimate its role in, for example, the Angola crisis in facilitating the insertion of friendly Cuban forces to sustain a newly established revolutionary regime.[15] Such activity remains conceivable in future, although there will be fewer opportunities to act given that there are no more traditional formal empires, except perhaps the Soviet Union's own, to break up. Moreover, the USSR does seem to have revised its approach to the utility of this degree of support for the Third World 'progressive' regimes. Countries sustained by Soviet and Cuban arms and aid have not proved necessarily to be loyal satellites. Even more importantly, the Soviet political leadership has come to recognize the dangers of pursuing the revolutionary process in an armed manner, except where absolutely unavoidable.[16] One may expect the Soviets to be even more reluctant than before in using its fleet. Nevertheless, in its world-wide deployment the Soviet Navy remains a vital factor in demonstrating direct Soviet interest in world events. As one study of

The international and legal context

Soviet naval diplomacy has put it: 'The Soviets are now universally understood to be fully fledged participants in the diplomacy of the Third World.'[17] The feathers of that plumage are in large part the ships and auxiliaries of the Soviet Navy.

The bird itself, however, must be sufficiently large and imposing if the feathers are to be of real meaning. As Edward Luttwak has made clear: 'It is the overall political context that will determine the political effectiveness of Super Power navies, and not their tactical capabilities, since it is this context that governs the application of national power, the true determinant of which the naval element is only one constituent.'[18] It is the overall power of the Soviet Union that is called into play when the Soviet Navy puts in an appearance. This makes up to a very considerable extent for any defects in capability on the spot.

Less powerful nations cannot hope to approach this level of peacetime naval diplomatic weight, even if they deploy navies of significant strength. Currently the two main ex-imperial powers, Britain and France, still deploy large and imposing navies, indeed the latter's gives a particular emphasis to operations outside Europe. From the mid-1970s, when France's carriers were redeployed to the Mediterranean, the French have been involved in a range of operations, notably off Lebanon, in support of their contribution to the multinational forces ashore. This involved two carrier air strikes, one against a Syrian battery and the other against a Palestinian camp. France's navy contributes to guaranteeing Djibouti's security and her Indian Ocean Squadron there allows her flexibly to operate in the Gulf and Red Sea as required by the perceived threat to French national interests and shipping. France has also quietly supported the Seychelles against South African mercenaries and has demonstrated resolve to protect her remaining possessions in the Indian Ocean, e.g. Mayotte from the Comoros, and Tromelin from Mauritius. France has similar requirements in the Pacific where her use of her Pacific islands as nuclear weapons testing sites has created a particularly sensitive context. Until the New Caledonia problem is settled the uneasy situation there is also one which will probably require continued and frequent naval deployments.[19]

France has chosen to integrate her far-flung relics of empire into the national state, something for which Britain has shown a traditional reluctance. Indeed, misleading signals of determination to hang on to the Falkland Islands tempted Argentina into an attempted island grab in 1982 that turned into the most notable naval event of the decade and the most significant traditional 'fleet versus fleet' battle of the post-war era. The war also showed that one should not underestimate the residual power of an ex-colonial state in possession

The evolving environment

of a few imperial relics. There is still a capability to embarrass those who go too far to exploit apparent weaknesses. Yet, a medium to long term view must share Sir James Cable's doubts about the long-term ability of those 'medium navies with ambitions above their economic station' in life to continue to play the *national* world power role.[20] When the case for a world-wide or 'out of area' role for a European maritime force is sustained by lists of past interventions, mostly carried out in a period when the state in question actually owned a long dismantled colonial empire or was following a responsible policy of making sure a pro-Western regime inherited the fruits of imperial withdrawal, the temptation to reach for one's intellectual pistol is especially acute. British naval spokesmen, both official and unofficial, were recently prone to such indulgences in imperial nostalgia, almost as if the sun had never set on a red coloured globe or the withdrawal from East of Suez in 1968–71 had never taken place.[21]

There are, however, two factors which may well provide extra substance to the 'out of area' pretensions of Europe's navies, both great and small. The first is the desire to maintain good relations with the United States. As former Secretary of Defence Caspar Weinberger made clear to his European colleagues at the Sealink Conference in Annapolis in 1986, some Americans, at least, feel the Europeans have a duty to support their American ally in dealing with the 'global maritime challenge'.[22] 'We must have leverage for action,' he argued, 'in corners of the globe far from the North Atlantic, places where developments can mightily affect NATO's security. We need to protect all of our NATO flanks. And we can only protect all of NATO and its flanks if we are fully prepared to take actions outside the NATO theatres. "Out of Area" can no longer be the automatic veto some feel that outworn geographical tag fastens on NATO.'[23] Weinberger went on to define as vital interests the 'critical sea lanes' in the Pacific, Indian Ocean, and the Gulf.[24]

The outbreak of war between Iraq and Iran in September 1980 had indeed seen British and French ships along with American patrolling off the entrance to the Gulf. These were purely national efforts on the part of the two European nations, although they did give comfort to the United States and were clearly appreciated. In 1987, shortly after a major British exercise in Oman, the naval dimension of the war, previously largely an Iraqi air offensive against Iranian shipping, was significantly stepped up. Increased Iranian attacks on ships in international waters and a build-up of Iranian capabilities to attack shipping serving Iraq's Arab allies, notably Kuwait, led to the American declarations that it would use force to maintain rights of passage for merchantmen in the Gulf.[25]

The international and legal context

During 1987 the British and French stepped up their own naval operations in the Gulf alongside those of the Americans, but were careful to maintain their respective efforts as national enterprises and within certain constraints. The British, for example, notably refrained from accompanying other than Red Ensign ships, and went no further north than Bahrain. During July the mining threat became serious but the British and French risked US displeasure, and showed their own reservations about certain aspects of US policy, by initially refusing to send MCM vessels to operate directly with the US forces. With, however, the extension of the mining threat on the south-eastern side of the Strait of Hormuz, in the middle of the Royal Navy's main operational area, the two European powers reversed their position in August and announced plans to send minehunters to operate in support of their existing forces. The British, indeed, also took the lead in trying to interest other European governments in a multinational MCM force.[26]

The Americans were very happy with this development. As a contemporary British press put it 'there was glee in Washington that, finally, one of the Western allies was doing something to ease the U.S.'s burdens in the Gulf'.[27] Initially, the other Europeans had reservations about joining the British but, assured of the military, technical and logistic support of British warships, the Dutch and Belgians decided in September to send out MCM vessels. Following an attack on one of their merchantmen the Italians went further and decided to send out a purely national full patrol of three MCM vessels and three frigates in order to give protection to their ships all the way to Kuwait.

The latter decisions had much more to do with national and European perceptions of interest than the desire to contribute to a US presence and respond to American pressure, or even control US action (although the latter may have been achieved). Moreover, the deployment of forces into a European multilateral framework played a vital part in the decision-making processes of some countries which were unused to sending their naval forces so far afield. Without such a framework, no deployment decision might have been possible. The Dutch used their tenure of the Chairmanship of the Western European Union (WEU) to bring together in August 1987 senior officials of all seven member states to discuss concerted naval action. There was little desire to set up a 'WEU Squadron'. The British, and even more so the West Germans, did not want to appear to be 'ganging up' on the Americans. France, pro-WEU at a general level, proved equally anxious to preserve freedom of action on the spot. She was, therefore, reluctant to enter into too integrated a force. The Italians, for their own political reasons, chose to emphasize independence also. Nevertheless, after the first August meeting a process of 'concertation' was established, by which the WEU framework was used as a forum for concerted action working at three levels: ministers and their senior representatives, experts, and the officers on the spot. Even Luxembourg gave financial support to the Dutch and Belgians, creating a Benelux flotilla which operated closely with the British, who provided surface ship cover. In 1988 the relationship was formalized and a joint Anglo-Benelux unit eventually formed under British command in order to allow a reduced presence of MCM vessels to be used to optimum effect. The WEU concertation process was used to achieve joint MCM operations in the summer of 1988 and also provided a context for wider rationalization. The West Germans helped cover for the absence of Italian and US ships from the Mediterranean by deploying forces there in late 1987 and early 1988. Finally, after the Gulf War ended, in the most obvious show of European concertation yet, the WEU took upon itself the role of clearing up the remaining mines in the southern Gulf Area in Operation Clean Sweep.[28]

The Europeans, almost despite themselves, crossed a number of thresholds of co-operation. The progressively closer Anglo-Benelux arrangements, first on mutual support and then on a joint force, were fundamental steps forward given previous doubts about even the legality of one navy's warships protecting another. The use of WEU to discuss co-operation was also a significant moment in the rebirth of this organization. The Europeans, dependent on Gulf oil for much of their supplies and interested in the freedom of navigation of their own vessels, were being drawn into a new kind of 'out of area' commitment, one that reflected real and modern necessities, rather than imperial echoes and one that was, slowly and diffidently, but nonetheless significantly extending the bounds of European defence and foreign policy co-operation. The European naval force in the Gulf might be a real foretaste of things to come as Europe's further integration inexorably continues and the United States' perceptions of the world may begin to diverge more and more as the latter looks West rather than East as its primary area of concern.

It has become increasingly fashionable to emphasize the growing strategic importance of the Pacific, both in itself and as a preoccupation for the United States. As Colin Gray told the audience at the 1986 Sealink Conference: 'We may debate the when and the how, but East Asia is becoming a – and perhaps even *the* – predominant center of world power. The security architecture of this region and the geostrategic significance of the region for other regions – NATO Europe in particular – will naturally change as China modernises and Japan adjusts to China's modernisation.'[29]

The international and legal context

Currently, Japan is the economic giant of the region, although she has yet to transfer that dominance into military or naval terms. Her 1945 aversion therapy to strategic ambitions remains strong and effective. Her heavy dependence on Gulf energy supplies has not shaken her traditional reticence about military involvement far from her shores. Japan shows every sign of continuing to be relatively cautious in adding military fledging to her new economic superpower capacity, but her navy will soon be deploying some of the world's most sophisticated non-superpower surface combatants. Given her extraordinary dependence on maritime trade and her possession to a greater degree than any other state of the traditional attributes of seapower, a large mercantile marine and a merchant shipbuilding infrastructure, it would be surprising indeed if her military potential at sea, and her willingness to use it if necessary, did not increase as World War II disappears increasingly into the mists of history. Japan's declaration of an interest in the defence of shipping a thousand miles from her coastline along the 'shipping lanes' to Taiwan and towards Iwo Jima is no doubt only a first step.[30]

It also seems likely that China will build up, albeit slowly, her maritime capabilities and her capacity to engage in naval diplomacy at increasing distances from her shores. The Chief of the Chinese Navy has stated that 'as the strategic position in the Pacific is becoming more important . . . and as China is gradually expanding the scale of her maritime development, the Chinese Navy will have to shoulder more and heavier tasks in both peacetime and war'.[31] The main confrontation is in the South China Sea with Vietnam and her Soviet ally, but from 1980 Chinese ships began to move further afield. In that year there was a naval cruise to the South West Pacific to recover rocket re-entry capsules, and in 1981 Chinese ships sailed round the Philippines. In 1984-5 Naval auxiliaries visited Chile and Argentina on their way back from establishing an Antarctic scientific station, and in 1985-6 a destroyer and an auxiliary cruised into the Indian Ocean to visit Bangladesh, Sri Lanka, and Pakistan. The latter deployment was of particular interest, especially as the small Chinese squadron exercised with the US Navy on the way home. One American analyst has summed up the significance of this cruise as follows:

> By Western standards the PRC naval cruise to the Indian Ocean was merely a routine training exercise. However, within the context of the Third World navies it was a significant accomplishment. Chinese sailors further proved their ability to refuel ships at sea and demonstrated their ability to maintain communications over great distances for an extended period of time. The political

The evolving environment

advantages accrued to Beijing were substantial and may prove to be of lasting benefit. Overall the cruise was a notable achievement in China's naval development and will probably be a stimulus for more distant visits over the next year or two.[32]

China's actual capability to project real military power much beyond the South China Sea will remain very limited for some time, but as her latent economic potential is unlocked there is little reason to doubt that she will use the growing reach of her navy as at least an expressive token of that strength and potential.

This potential growth of Chinese naval power acts as another stimulus to India to make her own the major Indian Ocean navy. China is not the only 'threat' however; India still remembers the US deployment of the *Enterprise* carrier battle group as an exercise in coercive diplomacy in the 1971 Indo-Pakistan War. With a navy centred around two light aircraft carriers (and with plans for a third of her own construction) operating modern STOVL fighters, a much more capable surface combatant fleet than China's, the beginnings of a nuclear submarine force and long-range maritime patrol aircraft, India possesses 'the largest and most balanced fleet in the Indian Ocean'.[33] She can lay claim to a respectable capacity to project military power into the seas around her shore, a capability that is more than one of mere defence. The main targets (or sometimes potential beneficiaries) of Indian naval power projection are her neighbours, Pakistan (the adversary above all), Bangladesh, Sri Lanka, the Maldives, Burma, and Indonesia. The main threat to Indian naval hegemony in the area will remain the presence of the navies of the superpowers, especially that of the United States. It is thus hardly surprising that India has been a major sponsor of 'zone of peace' proposals which might shut out outside actors and confirm her own naval hegemony. Until such measures come into effect, India will seek to use her navy as a deterrent to the projection of power into the Indian Ocean from outside. India's powerful navy, however, might be a two-edged sword. As one informed analyst has warned:

> In so far as it complicates the force planning of a particular extra regional threat it will have served the defensive purposes for which it was created. Its inherent capability to unnerve India's regional neighbours, however, is a political issue that merits greater attention. Unless India's military expansion is accompanied by a greater willingness to be more sensitive to the concerns of smaller neighbours, its force superiority will never adequately translate into political pre-eminence. In these circumstances, the naval forces acquired for the general purpose of greatness may

become the very source of larger political discomfort and discord.[34]

An Indian Ocean 'zone of peace' measure may indeed be a contribution to the solution of this thorny problem. This could legitimize some significant 'peacekeeping' role for India in the area which is clearly acceptable for countries like Sri Lanka and the Maldives, which have already accepted Indian intervention. Such a role, however, would be clearly unacceptable for Pakistan unless Indian ambitions and capabilities were themselves significantly curtailed. Other, more distant Indian Ocean powers, like South Africa and Australia, would also be concerned to see India's wings clipped by such a measure.

A world where superpower rivalry has had the edge taken off it by a new and more substantial era of *détente*, and where other naval powers, including Europe in new political forms, are becoming more active, also allows other states to play a more leading role in the local naval balance. They may indeed be able to assert a local naval leadership. Such a state is Australia, currently reassessing her defence policy in a more local context and de-emphasizing her traditional role as an expeditionary contributor to British or American campaigns. Although Australia has not abandoned her commitments on the Asian mainland, she has been taking a careful look at the defence priorities of 'one of the most secure countries in the world ... distant from the main centres of global military confrontation, and ... surrounded by large expanses of water which make it difficult to attack', especially a country with neighbours possessing 'only limited capabilities to project military power against it'.[35] In his wide-ranging report presented to the Australian government, Paul Dibb identified as an important security interest 'a strong stable region free from external pressures'. The area for exerting 'independent power' was defined as stretching 'over 4,000 nautical miles from the Cocos Islands in the west to New Zealand and the islands of the South West Pacific in the east, and over 3,000 nautical miles from the archipelago and island chain in the north to the Southern Ocean. It represents about 10% of the earth's surface'.[36] And most of it is sea.

Dibb went on to define Australia's national role as follows: 'In the South Pacific we are perceived as being by far the largest power. Our fundamental security interest is to maintain the benign strategic environment that currently prevails, free from unwelcome external pressures. Our foreign policy, aid programmes, and defence policy should be co-ordinated carefully with other regional states, in particular New Zealand, so as to discourage Soviet naval visits, or other unwelcome military access in the South Pacific.' In other words, local actors of significant power prevent outsiders taking advantage of power vacua.

This policy of 'maintenance'[37] implies powerful maritime forces, at least by local standards. Land-based, over-the-horizon radars and aircraft, along with conventional submarines, provide useful surveillance and strike capabilities as well as defensive air cover, but reach and presence can only be provided by a surface fleet. Dibb recommended twenty-four 'platforms', a three-tier force of up to nine destroyers (currently American guided-missile destroyers and large frigates), and up to ten 'light patrol frigates' with a similar number of smaller patrol vessels in the 200 ton class.[38] Preliminary work on the new frigates is being pursued jointly with New Zealand. The latter, currently isolated from the USA because of its anti-nuclear rhetoric, places much emphasis on its continued defence connections with Australia at the various levels within the Australian/New Zealand Defence Consultative Committee (ANZDCC) framework. Such consultations reflect New Zealand's recognition that 'it shares vital strategic interests in the South Pacific with Australia'.[39]

Given the limited opposition in Southern Pacific waters, developed countries like Australia and New Zealand are regional powers of some importance with a long reach measured in thousands of miles. In many other areas, however, albeit over shorter distances, local powers will feel the need to assert their sovereignty and control in areas for which they feel responsible, and in which they feel legitimately interested. This may well cross the line between mere constabulary duties and 'gunboat diplomacy'. As Sir James Cable has put it: 'Because more nations than ever before now have sea coasts, navies and maritime interests to quarrel about, political considerations make it likely that threats and/or limited uses of naval force will grow rather than decline during the rest of the century.'[40]

What seems to be emerging is a system where the superpowers will remain for a time supreme in the area of what Luttwak calls 'naval suasion'.[41] They, especially the Americans, will use navies to compel smaller nations to do something, to make them desist from some activity, or to deter them from carrying out certain actions contrary to the perceived interests of themselves and their clients. A second group of powers, some on the upward slope, some on the downward, will attempt with more difficulty to act as assailants in exercises of limited naval force. Those on the downward slope may, however, play a significant role in superpower confrontations, and they may also learn the usefulness of acting collectively. A mass of medium and small states will always have the motivation and means to defend or assert their national interests at sea either collectively or, more usually, nationally. This is especially so given the connected developments in both resource exploitation and the law of the sea that give all littoral states substantially increased off-shore areas of responsibility and control.

As the sea has been used more it has been more necessary to regulate its use through the medium of international law. The attempt to create a universally accepted modern legal framework that was the Third United Nation's Conference on the Law of the Sea (UNCLOS III) has not yet been fully successful. Despite eight years of negotiation, from 1974 onwards, complete agreement proved impossible on one outstanding issue, the future of the deep seabed. President Reagan's USA loyally followed by its two most faithful allies, Britain and the German Federal Republic, refused to sign the resulting United Nations Convention on the Law of the Sea that opened for signature at Montego Bay in 1982. Although no less than 159 countries had signed the Convention by the deadline of December 1984, the opposition of the Americans, British, and West Germans continued.[42]

This hostile attitude still prevents the 1982 Convention from forming the formal basis for the international law of the sea. Nevertheless, its main provisions, which were already becoming the norm anyway – a twelve mile territorial sea over which littoral states have full sovereignty (saving rights of other's innocent passage), a 200-mile exclusive economic zone in which coastal states have the sole right of exploitation of living and non-living resources, and similar monopoly rights in certain circumstances to manage the same resources on adjacent continental shelf beyond that – have become accepted as the customary law of the sea. Non-signatory nations, like the United Kingdom, have extended their territorial seas to twelve miles, and it will not be long before all states have unilaterally conformed to the new rules, both in the rights they assert and others' rights they accept.

Even the 1982 Convention's greatest enemy, the United States, has stated that most of its provisions 'reflected prevailing international practice' and that, therefore, 'their substance . . . could be invoked by non-parties as representing new customary international law'.[43] As Ken Booth has pointed out: 'The history of the law of the sea has been predominantly a story of the evolution of customary rules through state practice. The factors contributing to order at sea have always been relatively strong.'[44]

Nevertheless, especially in a context where there is no universal signed and sealed agreement on what the law of the sea actually is,

The evolving environment

there is still scope for conflict. In his excellent paper, 'A Sea of Troubles',[45] Barry Buzan analysed the 'sources of dispute' that could arise over law of the seas questions. These came in four groups, one concerning disagreement *over* national boundaries, the second concerning the rights *within* national boundaries, the third over 'rights in the ocean beyond national jurisdiction', and the fourth 'disputes arising from non-ocean sources'. Among the first are (a) 'delimitation of opposite or adjacent boundaries', (b) 'ownership of, or rights pertaining to, islands', and (c) 'delimitation of the outer limits of national jurisdiction'. The second includes disputes over (a) fishing rights, (b) non-military navigation rights, (c) rights to carry out military activities, (d) rights to carry out marine scientific research, and (e) rights to exploit the seabed. The third general category is not subdivided and it has indeed turned out to be *the* major problem in preventing universal adoption of an agreed law of the sea regime. The attempt to make the distant oceans 'the common heritage of mankind' brings us to the fourth major category which includes disputes over (a) 'the rights of land-locked states', (b) 'the distribution of power or rights within states', and (c) 'territorial disputes involving coastal states'.[46]

In viewing the 'law of the sea' problems that have arisen during the 1980s one can see examples of several of these categories, often in an overlapping form. Disputes between the Soviet Union and its Scandinavian neighbours over maritime boundaries led to the cutting of a Norwegian seismic research ship's cable and the boarding and expulsion of various Swedish, Danish, and West German fishing boats in the Eastern Baltic. The United States deliberately exerted its right to sail warships through territorial sea limits it did not accept and to exert the right of innocent passage of warships in territorial waters over the protests of the sovereign power. (Sweden and the Soviet Union were the respective targets.) The conflict over Libya's claim to internal waters sovereignty over the Gulf of Sibra came to blows in 1986, as it had in 1981. The United Kingdom very successfully established a 150-mile fishery protection zone around the Falkland Islands which directly conflicted with Argentina's 200-mile Exclusive Economic Zone (EEZ). Canada asserted internal waters claims to the North West Passage which the United States had opposed to the extent of sailing a Coast Guard icebreaker through the disputed area without asking Canadian permission. The uninhabited islet of Rockall and its associated continental shelf rights remained a bone of contention between the United Kingdom, Ireland, Iceland, and Denmark. France had problems with the USSR over seabed exploitation rights, and asserted claims to forbid research in its EEZ except by permit, while Norway prevented the Greenpeace

The international and legal context

ship *Moby Dick* entering its territorial sea without permission.

One should not overstate the significance of legal disputes such as these. Where there is a political will, peaceful settlement is both possible and probable. One notable case is the Beagle Channel settlement between Argentina and Chile ratified in 1985. Thanks to Vatican mediation the dispute, which had caused considerable tension between the two states concerned, was settled on the basis of the sovereignty of three islands going to Chile with an adjacent twelve miles of territorial waters, but Argentina retaining a 30–40 mile EEZ in the Atlantic. The International Court of Justice proved useful in arbitrating other disputes, e.g. between Tunisia and Libya, and Malta and Libya. Indeed, only where wider political factors created additional tensions, notably in the US–Libyan affair, was conflict pushed to the level of armed confrontation involving the two sides' naval forces.[48] The wider political factors were clearly more important. The United States, so anxious to uphold the law against Libya, totally ignored the strictures of the International Court of Justice when it was criticized for laying mines in Nicaraguan territorial waters. A Nicaraguan Security Council motion to enforce the judgement was simply vetoed by the United States.

As usual in international politics, international law is only useful when two states wish to settle disputes by agreement. These, happily, cover the majority of cases but some have been disturbed by 'a broader willingness on the part of the major powers to disregard the territorial sovereignty of other states and, as a matter of deliberate policy to place military or police expediency above compliance with international law'.[49] The superpowers are major offenders. The USA's action in using carrier aircraft to seize an airliner carrying the *Achille Lauro* hijackers was one such example, along with the reported violations of the territorial waters of Sweden and other neighbouring states by Soviet submarines of various shapes and sizes. This is a dangerous and slippery slope. Other aspects of this dangerous trend to lawlessness are the continued prevalence of piracy and the beginnings of maritime terrorism, and the equally lawless great power responses to it just discussed (see Chapter 10).

Nevertheless, one should not be too pessimistic. It is likely that the dominant trend will in fact be the opposite, towards an increasingly firm framework of maritime law at sea. As the disadvantages of non-compliance with the UN Convention became clear, e.g. the non-access to UN procedures for settling seabed exploitation rights (as agreed within the framework of the 1982 Convention by the USSR and Japan), the small rejectionist front might crack. Whatever the precise status of the 1982 Convention, however, as Ken Booth has pointed out, UNCLOS III might well not 'be the end of the conference circuit but merely the first lap'.[50]

This brings us to the phenomenon of Creeping Jurisdiction, defined by Booth as 'the extension of national or international rules and regulations, and rights and duties over and under the sea, in straits and coastal zones, on and under the seabed and in the vast stretches of the high seas'.[51] It is hard not to agree with Booth when he goes on to argue that 'In future, technology, interest and the will to govern seem set to fill out large chunks of the map of the sea with appropriate forms of national and international administration as inexorably as railways, the industrial revolution and nationalism spread control throughout the land masses in the last century.'[51]

The exact nature of the EEZ jurisdiction has already led to disagreements, and these will continue. Moreover, foreign naval forces may well come in for unwelcome attention and attempted restraint. Booth concludes that

> the secular trend is adverse to the naval interest – as traditionally and conventionally understood – despite the rights enshrined in the 1982 Convention . . . the long term future of naval mobility is under challenge. It may take half a century but unilateralist drives to parcel parts of the ocean will continue and will be legitimised by the territorialist mood of the international community. As this development unfolds, and as state control intensifies over larger patches of sea, greater meaning will be invested in the new boundaries which are inevitably the outcome of the process. Nations will feel protective and sensitive – indeed patriotic – about these patches of ocean. Hence it is appropriate to talk about the spread of what might be called 'psycho-legal boundaries' at sea.[52]

The major naval powers will inevitably oppose the creep of 'territorialisation and psycho-legal boundaries' but analysts such as Booth and Alexander[53] put forward convincing cases that this will be a rearguard action. Increasingly, legal questions will interact with political ones in exercises of naval power, especially at the lower levels. Indeed, creeping jurisdiction will create naval opportunities as well as problems. Having boundaries that are such but which are 'less immediately sensitive and further from the national nerve endings than those on land'[54] could well provide new and useful options for assertive and forceful activity without the same risks of similar confrontations on land.

The greater territorialization of the oceans may well act as an extra impulse to explore more international uses of naval power. As the relations of the two superpowers improve, the barriers to

The international and legal context

international peacekeeping may well decline. Soviet 'new thinking' in the mid-1980s has dramatically changed Soviet approaches to UN peacekeeping in a positive direction. Indeed, it was the West that strongly opposed a UN naval force being set up in the Gulf. Although there were other more practical problems, some in the West feared such a force might legitimize the existing Soviet presence in an area of very significant Western interest.

Such concerns, however, may diminish in the future. As they do, the technical and legal problems that have traditionally been held to prevent UN maritime action – problems of mutual defence if acting outside Chapter 7 of the Charter, problems of common communications and rules of engagement, problems of command – might well suddenly seem soluble. Indeed, the more general possession of sophisticated naval capabilities by non-aligned countries such as India might allow a wider spectrum of potential participants in joint peacekeeping operations. There are useful operational precedents for joint action by a large number of different navies in a UN framework, e.g. the Korean War.[55]

Whatever the political framework, the evolving political and legal context will allow wide scope for the use of maritime military power. It would be fair to allow Ken Booth, upon whose work this chapter has relied so heavily, to be allowed the last word, especially as it helps clear up a slightly naughty case of a quotation out of context. James Cable took Booth to task in a recent article for implying that traditional 'gunboat diplomacy' was dead.[56] The full statement reads as follows and sums up the findings of this chapter very well: 'While there is no prospect of the revival of gunboat diplomacy which characterised the age of imperialism in the last century', naval forces 'will remain instruments of diplomacy for several countries and the defenders of the maritime sovereignty of many more'.[57]

Chapter 8

Economic stimuli and constraints

Mahan argued that military sea power grew out of the economic uses of the sea and the importance of those uses to the state. The seas, however, are now arguably less central to the power position of the great powers than they once were. The modern Mahanian can quote persuasive figures of greatly expanding world trade volumes, more than the lion's share of which goes by sea (see Chapter 3). Certain countries are indeed more dependent on overseas trade than they once were, e.g. Japan which exported about 10 per cent of its GNP in 1960 and exported about 14 per cent a quarter of a century later. Japan also depends entirely on seaborne trade to keep all her economic activities going in terms of energy and raw materials. Britain still exports about 20 per cent of her GNP and depends on the seabed for the energy supplies of oil and gas that are vital for her agriculture as well as her industries. Other states for whom seaborne exports are as important as they are to Britain and Japan are states such as Israel, The Netherlands, New Zealand, Norway, Nigeria, Sweden, and Venezuela (all 20 per cent or over), and Argentina, Australia, Canada, Chile, Egypt, France, the German Federal Republic, Iran, Italy, Saudi Arabia, and South Africa. Oil-producing states are especially dependent on seaborne trade and oil is also a particular item of export dependence in the developed world. It is no coincidence that The Netherlands, dragged somewhat against her will into a far-flung naval commitment, is almost two-thirds dependent on Gulf supplies of oil.[1]

The continued economic interest of such states in using the seas is clear, although maritime policy is shaped in interesting ways by extraneous factors. In Japan's case, for example, her post-1945 reticence about exerting military power to parallel her economic strength has limited her military navy in both size and capability. These restraints are steadily and slowly loosening, but it will be a long process. In the United Kingdom's case the civilian aspects of sea use have been allowed to wither to a remarkable degree. British governments, especially the current one with its fierce ideological commitment to the free market, both national and international, leave Britain's maritime industries to be squeezed by states with less liberal ideals and practice and/or less expensive ships and crews.[2] Britain's military naval forces also have become primarily orientated to power projection in both its nuclear and conventional forms as they have been given both the custody of the national deterrent and become locked into the United States' side of the superpower balance.

Neither major naval superpower relies on the economic uses of the sea. The United States exports between 4 and 5 per cent of its GNP, and over a quarter of those exports go to Canada and Mexico. Given the USSR's connection by land to her major markets her figure is even lower, perhaps about 3 per cent, although her merchant marine is a significant 'export' in itself, especially as a hard currency earner. The Soviet Union is, however, self-sufficient in terms of dependence on overseas imports. She has to import significant amounts of food products to sustain domestic living standards but she does not rely on overseas supplies. She is also very well off for domestic supplies of oil (with the largest reserves outside the Gulf) and non-fuel minerals, although she does import significant quantities of the latter from non-CMEA countries, especially tungsten and bauxite/aluminium.

With the United States the matter of import reliance is much more controversial. Much is often made of American 'dependence' on overseas supplies of fuel and raw materials. As President Reagan has put it: 'The United States is a sea power by necessity, crucially dependent on the trans-oceanic import of vital strategic materials. For this reason our navy is designed to keep open the sea lanes world wide . . . maritime superiority is for us a necessity.'[3] The figures to sustain this theory look superficially impressive. The United States still imports about 30 per cent of its oil, and is heavily dependent on imports for a wide range of non-fuel minerals, many of crucial industrial importance, such as manganese, cobalt, and chromium. For all these substances import dependence is 90 per cent or over and the suppliers are all overseas.[4] Understandably, American navalists make much of these figures but the real situation is rather more subtle. The United States itself is well supplied with deposits of all minerals but chooses not to mine many of them because cheaper supplies are available elsewhere. US vulnerability to a mineral cut-off is often overstated. As one attempt to put the other side of the picture has put it: 'A combination of "safe" production, substitution, domestic production, conservation and recycling could readily compensate for a loss of imports. The few minerals for

The evolving environment

which this is not the case are non essential.'[5] Moreover, the USA maintain significant stockpiles of critical materials, enough to deal with short- to medium-term disruptions.

Neither superpower is therefore a small island critically dependent economically on the use of the sea as the British Empire was at the height of its power. Both are essentially continental superpowers. The United States is linked into a complex, international, and interdependent capitalist economic system where it chooses to buy in the cheapest markets rather than face the cost in living standards and prosperity dictated by a self-imposed blockade. The USSR operates as the centre of an alternative economic world, but one increasingly conscious of its need to interact with the rest of the globe economically and politically. Both, therefore, rely on the sea as a matter of economic *well-being* rather than *survival*. The one's use of the sea is in this sense no more or less 'legitimate' than the other's.

It must be stated, however, that in any kind of protracted conventional conflict the United States would require substantial imports of raw materials from overseas to exert its maximum strength. Moreover, the USA has economically related reasons to deploy more military force at sea than its Soviet adversary partner; the OECD world as a whole is more dependent on sea communications than CMEA (Comecon). The former contains, as we have seen, a highly sea-dependent island medium power and an economic giant, weakened in military terms. Part of the bargain that the allies of the United States make with their patron includes an American commitment to maintain freedom of navigation by force if need be through the 'maritime superiority' of which President Reagan spoke above. This 'bargain', however, is far from being a straightforward one. Legally, the United States cannot even protect directly that majority of US-owned ships that does not fly the stars and stripes. Moreover, the diminishing relative economic power of the USA *vis-à-vis* the rest of the Western world argues for some changes in the relative share in the provision of defence for the West's maritime trade. Those states that are more sea dependent may well have to do more for themselves, either individually or collectively.

Two tables of the relative economic interest that states have in using the sea have been prepared, one by Alistair Couper and his Cardiff University team for the *Times Atlas of the Sea*, and the other by Admiral Richard Hill for his excellent study *Maritime Strategy for Medium Powers*. The Cardiff University team were only concerned with interest in seaborne *trade* and took three criteria: (1) total seaborne trade import and export, (2) seaborne trade as a percentage of GNP, and (3) size of merchant fleet. Different indices were

Economic stimuli and constraints

prepared for parameter (3), one based on beneficial ownership, the other on flag of registry. On the former criteria the indices were as follows. The data upon which calculations were based were for 1980.[6]

Interest index
9 – 10 Japan
8 – 9 Saudi Arabia
7 – 8 Kuwait, The Netherlands, USA, United Kingdom, Singapore
6 – 7 Italy, Norway, Libya, Iran, Greece, France
5 – 6 Venezuela, Spain, USSR
4 – 5 West Germany, Australia, Algeria, Qatar, Canada, Denmark, Brazil, Sweden
3 – 4 Belgium, Nigeria, India
2 – 3 Poland, Liberia
1 – 2 Tunisia, Gabon
0 – 1 Lebanon, Syria, Panama, Trinidad and Tobago

Adjusting parameter (3) for flag registry produced some interesting changes.

Interest index
9 – 10 Liberia, Japan
8 – 9 Saudi Arabia
7 – 8 Kuwait, The Netherlands, United Kingdom, Singapore
6 – 7 Italy, Norway, Libya, Iran, USA, France
5 – 6 Greece, Venezuela, Spain, USSR
4 – 5 Australia, Algeria, Qatar, Denmark, West Germany, Brazil, Sweden, Canada
3 – 4 Nigeria, India, Panama
2 – 3 Poland
1 – 2 Tunisia, Gabon
0 – 1 Lebanon, Syria, Trinidad and Tobago

The Hill indices brought in both the question of *dependence* by relating degree of interest to GDP and population, and broadened the scope of the study to cover shipbuilding, fish catches, and off-shore zones. The data is now a decade out of date but the general results probably still have some validity. On the basis of GDP, Norway emerged as the most sea dependent nation by a substantial margin, followed by Denmark and Argentina and then Japan, Greece, and Brazil. Next came Chile, The Netherlands, and Spain leading a large group of Australia, Egypt, Finland, Iran, Italy, Nigeria, Sweden, and the United Kingdom, all at about the same level. Canada and India led the next group with Belgium, France, Mexico, New Zealand, South Africa, and Venezuela all about on a par with the USA. Finally, there came West Germany and Israel.[7]

When dependence on maritime economic activities was compared with population the results were a little different. Norway still led the table, but Denmark and Japan were closer behind than they were on the basis of GDP, as were The Netherlands, Spain, Sweden and the United Kingdom, and then Canada, Australia and Finland. France then led the United States and Belgium followed by Italy, Argentina, Greece, Chile, Venezuela, South Africa, West Germany, and New Zealand. Then came Mexico with a gap before Brazil which led India, Iran and Israel, and finally Nigeria and Egypt. The spread of degrees of dependence seemed to be larger on a basis of population than it was on that of GDP.[8]

Admiral Hill sums up his 'unspectacular but useful' results as follows:

> The most consistently sea dependent nations are either islands or states with long coastlines and limited land frontiers. Of those in the tables Japan, the Scandinavian countries and the United Kingdom stand out; the pattern is repeated for smaller states like Singapore, and probably would be for a state like Cuba if its figures were available. Other nations may be less consistently sea dependent but still have a very marked dependence on one aspect or other of the sea affair. Liberia on merchant shipping, Saudi Arabia and Kuwait on shipborne trade, Iceland on fishing, Gabon on the offshore zone; for all these dependence is extremely marked and . . . their importance to the world economic and strategic situation makes their vulnerability of some significance.[9]

Certainly almost all the states mentioned above would seem to have some interest in deploying maritime forces of some kind, although there is no direct and necessary connection between the extent of dependence and the degree of capability of those forces. Open registry states do not, on the whole, feel a responsibility to defend their merchantmen. One can, however, protect ship*building* by ordering warships; and offshore zones always need policing. The latter factor has probably fostered more naval improvement than any other in recent years. There is thus still a connection between economic dependence and interest and military naval capability, although the reach of a maritime nation's maritime forces reflects both a country's absolute level of economic power and its perception of its political role more than it does its degree of sea dependence.

In some ways sea dependence is steadily increasing. The decline

Economic stimuli and constraints

of domestic heavy industries in Europe and the USA has led to a remarkable degree of import dependence for certain items, e.g. motor vehicles. Most important of all, however, is the dependence on imported energy of much of the developed world. In the late nineteenth and early twentieth century the realization that Britain was dependent on overseas supplies of food acted as a potent spur to naval expansion. Now Europe has more food than it can eat. Nevertheless, import dependence has only been pushed back one remove: modern farming techniques are heavily dependent on imports of energy.

The most important single component of maritime trade is, as we have seen in Chapter 2, crude oil and petroleum products. H.W. Maull in his major study, estimated that 'the industrialised countries will remain dependent on oil imports, although in terms of absolute quantities there will be little, if any, increases and perhaps even a reduction'.[10] Indeed, import dependence may well actually increase, with the Middle East asserting a more important position as the main source.[11] The oil nexus may well draw the producing states into increased naval roles just as it will emphasize the need for sea control with long reach in the hands of the importers. It is thus hardly surprising the the Gulf has already been the context for the most significant innovations in naval operations in defence of shipping, i.e. re-flagging and, more importantly, European co-operation.

In the longer term, into the next century, the oil trade may go into absolute decline as reserves are depleted and the prospect of more expensive oil stimulates synthetic production. Alternatives, like natural gas and coal, will become more important. The former sometimes comes from the sea, but the main producer is the USSR with its large underground fields. Gas also lends itself to long pipeline transit, both overland and across short stretches of the seabed (e.g. from Algeria to Western Europe). Japan will, however, continue to rely on sea transport for its natural gas supplies.[12]

Coal may well become an important cargo. It is currently shipped in large quantities by sea and will be transported in even greater quantities over longer distances in ships from the United States, Africa, China and other East Asian states to countries, like those of Western Europe, less well endowed with cheap, easily exploited seams. The next century may well be the age of the 'super collier'.

As oil becomes more expensive, and the exploitation of undersea oilfields becomes more attractive once more, the scope for maritime conflict over access to the seabed may increase. This could be especially true in Asian waters, where there remain serious disputes over the ownership of various archipelagos sitting above undersea oilfields. 'Very few islands in the South China Sea are

177

uncontested.'[13] Both China and Vietnam claim the Paracel (or Hsi-Sha) Islands, occupied forcefully by the former in 1974, Taiwan and the Philippines claim an interest in these islands also.

The Paracels are also significant in terms of fishing. Demand for fish is expected to rise to between 113–125 million tonnes by the end of the century. As the 'sustainable harvest' is estimated to be no more than 100 million tonnes the excess demand for fish may well fuel disputes over management of stocks and access to fishing grounds.[14] Where exclusive economic zones are not currently properly policed there will be increased efforts to maintain rights to control and license exploitation of national resources. Asian waters (including the Indian Ocean) will once again be a focus for this activity.

Economic pressures may well thus militate in favour of disorder at sea, and an enhanced role for naval forces in both their military and law enforcement roles. Nevertheless, one need not be too pessimistic. Both in terms of oil exploration and fishing an agreed 'pax' is necessary before the seas can be economically exploited. Oil men in particular are easily scared away by their inherently expensive and dangerous activities at sea being thrown into even more jeopardy by human opposition. This will act as a potent factor to achieve peaceful settlements, albeit settlements that will have to be backed up by continuous patrols by maritime law enforcement forces.

One maritime industry that has not yet been mentioned as a stimulus to naval power is shipbuilding. This industry has naturally suffered from the supply of shipping exceeding demand and has been in heavy recession, although signs of an upturn were noted in 1987. The most important shipbuilders, accounting for over three-quarters of world output, are Japan and South Korea, virtually first equal with about a quarter of world tonnage each, with the other quarter divided between Yugoslavia, Italy, Brazil, Poland, China, Romania, and Taiwan. Subsidies help to maintain Italy's shipbuilding industries, and Poland had been sustained by orders from the USSR in the non-capitalist CMEA framework: few successful shipbuilders today operate a fully commercial environment.[15] This means that non-commercial shipping, i.e. warships and support vessels, are often the key to the survival of any significant shipbuilding industry in certain countries, notably in the UK and the USA, which had *no* major merchant ship under construction in domestic yards at the beginning of 1988. Shipbuilding is secure only by becoming part of the defence industries that such states sustain for security reasons. The specialist skills required are still assets that give a competitive edge to states uncompetitive in the construction of simple merchantmen. Even free

Economic stimuli and constraints

market governments can feel happy about using naval shipbuilding as a form of economic relief, as long as there is an overall national security interest in obtaining the vessel in question. Thus can economic factors become a factor in sustaining naval strength.

As the sea gets more important in various ways this may have an effect on the internal political debate in various states that will make governments more sea minded and therefore more likely to fund maritime forces. There are, however, countervailing factors. The traditional industrial nations (and sea powers) seem to be moving away from production and transport. As Admiral Sir James Eberle pointed out to the Sealink Conference in 1986: 'International financial flows, rather than trade flows, now form the dominant part of international activity. Despite increases in the volume of international trade and its importance to manufacturing and industry only some 2 per cent of international money transactions now relate to payments for international trade.'[16] Electronic connections are the arteries of much modern commerce, not ships. If financial connections are the important factors then ownership of fleets and the wherewithal to protect them may not seem important any more. As Sir James Cable has pointed out with reference to the United Kingdom, 'the complacent regard merchant shipping as a primary industry that can safely be abandoned to the Third World as long as the vessels are still insured at Lloyds'.[17] One can sympathize up to a point with the nationalistic criticism implied by the above statement. Certainly, an increasingly internationalized world shipping industry is coexisting increasingly uneasily with the established naval and legal maritime order based around national flags and national navies. As Sir James Eberle succinctly put it to the delegates in Annapolis: 'If you sink a ship owned in Britain, financed in the Europe market, registered in the Bahamas, crewed from Hong Kong, insured in London, carrying Japanese cars whose engines were assembled in Germany, whose interests are you mainly damaging?'[18] The answer is many peoples', most of whom will expect someone, but not necessarily themselves, to sort the situation out.

The world in general, and the seas in particular, have never been so internationalized, a reflection of a wider economic phenomenon: 'Modern industrial societies, and indeed the present world economy, are built on reliance and interdependence, on free flows of goods, capital and technology and thus generally on the exploitation of benefits which can be reaped from the international division of labour.'[19] As this international division of labour becomes ever more dispersed, however, the provision of naval protection for its maritime dimensions is going to have to be reconsidered. Warships flying one flag are going to have to become much more flexible in their attitudes to protecting shipping flying another. This can be done by national decision; both France and the USA announced that it would defend all Gulf shipping regardless of flag in January 1988;[20] but perhaps a resurrection of the UN political framework for such activities might help these politico-legal problems to be fully solved on a more permanent basis. Sadly, however, the decline of *national* maritime industries is likely to lead to a widening of a phenomenon noted by Richard Hill in his typically pungent style: it is a 'malign neglect that can most kindly be characterised as sea blindness'.[21] This could have serious negative effects on naval capabilities, especially as warships are not getting any cheaper.

This brings us to economic constraints on naval capabilities. No navy is free of such problems, least of all the world's largest, the United States Navy. Although in some ways the massive budgetary deficits run up by the Reagan Administration in the 1980s have been a novel and creative way of mobilizing Japanese and Western European financial resources to pay for US naval power, the consensus view both nationally in the United States and internationally in the world financial community is that 'the deficit' should be cut. Of the two influences, domestic US opinion is by far the most important. 'Americans are convinced that federal spending generally – and defence spending specifically – must be brought under control.'[22] The result of this conviction, the Gramm-Rudman-Hollings Act of 1985 has created a new environment in Washington, 'forcing the hard questions' on defence spending.[23]

In a context where a 3 per cent increase in real terms was required to maintain the 600 ship/15 carrier battle group fleet that emerged from the John Lehman era, funds began to be cut in real terms from after 1985. In 1987 one sympathetic observer expressed 'great concern that Gramm Rudman and its related trends will make it impossible for the Navy to meet all of its requirements, especially as the competition intensifies with the other services for increasingly scarce funds'.[24] Sure enough, February 1988 saw the 600-ship navy target falter, with the decision to cut the American frigate force and even consideration being given to the possible paying off of the three oldest carriers.[25]

It will have been noted above that *increasing* expenditure was required to maintain US naval strength. 'Level funding' does not solve a navy's financial problems, as the current British Royal Navy finds. Naval costs have an inexorable tendency to spiral upwards under inflationary pressures all of their own. These have been analysed most recently by Philip Pugh.[26] He points out that increasing capability and increasing cost are an inevitable result of trying to keep up in the military competition by increasing capabilities to match those of an opponent. Naval power, depending on expensive platforms like ships and aircraft, is especially vulnerable to such cost escalation. He also demonstrates that ever since industrialization unleashed the prometheus of technological development, naval budgets have been tending towards the 'absurdity' of only being able to afford one ship. The British naval budget from 1860 to 1885 and the American naval budget from 1960 to 1985 both display this frightening characteristic. If the lines of unit costs and new construction budgets are extrapolated for two decades or so they cross. This absurdity, however, was avoided through a mix of three measures: increasing the budget thanks to economic growth, a reduction in numbers of units, and alterations of structure towards the procurement of smaller and cheaper vessels.

Pugh bases his case on the post-1950 Royal Navy, a fleet specially afflicted by the conflicting requirements of force enhancement and inadequate resources. Yet it seems that other navies less prone to pressures to live beyond their means have responded to the same stimuli in similar ways. 'Escalating costs,' Pugh argues, 'have demanded change towards smaller navies equipped with smaller ships; but that has not of itself meant any change in the balance of naval power. Alterations in that follow from changes in the balance of defence budgets and, hence, from differences in the economic performances of nations.'[27]

Optimizing the naval budget, however, as Pugh goes on to point out, is a tricky business. The natural tendency to try to exploit new technology to the full leads to ever higher development costs and another potential *reductio ad absurdum*, all money going on development and none on actual procurement. Spacing out development programmes and updating existing weapons limit development costs in their different ways, especially for missiles and aircraft that incur especially high development costs. It is indeed surprising how long lived *basic* missile systems are in an age of rapid technological progress. In some cases their capabilities have been improved, sometimes transformed, by relatively small changes to the basic airframe.

Another balance that has to be struck is that between individual ship capabilities and the quantity of units required. This dilemma operates both within the same class of ships and between different classes and sizes of vessel. This is a most difficult balance to achieve, especially as smaller vessels may not necessarily be the more cost effective. Smaller vessels may indeed by *less* efficient, given the higher relative costs of the hull relative to the weapons systems carried. A 6,000 ton combatant devotes about 12 per cent of its cost to floating, 14 per cent to moving, and 74 per cent to

The evolving environment

fighting. A 750-ton vessel devotes 13 per cent to floating, 29 per cent to moving, and only 58 per cent to fighting.[28] Thus 'restraining the size of warships is not a wise policy unless there are overriding reasons for doing so. . . . Reasons for setting a lower limit to the number of ships are: a recognition that no vessel, however good, can be in two places at once or, else, anxiety to ensure that accident or misjudgment cannot deprive a fleet of too great a proportion of its strength.' Pugh argues convincingly that it is 'fundamentally mistaken' to make

> small size and/or low unit cost into a primary goal of procurement policy . . . if halting the rise in unit cost is made the first priority then costs will cease to rise. But the penalty of such an achievement will be a disproportionate loss in capability and hence worse value for money spent on the fleet as its formation and the specification of its ships drifts further from their optimum . . . the rise in unit costs is not inevitable; but it is unavoidable if limited budgets are to be spent rationally and to the best effect. Sailing with this tide has obtained the best value for money that was to be had. If, as a result, cherished roles and missions have had to be given up then the harsh logic of the situation is that attempts to retain them could only have done greater injury to the overall capabilities of the navy in question.[29]

Navies cannot just keep their existing vessels forever, given both the fact that machinery wears out and manpower is an ever more expensive resource. The latter factor may even prevent continued procurement of new examples of the same designs. Navies usually lack men more than they lack ships or aircraft.[30] Trained manpower of sufficient quality for modern naval operations is a scarce commodity the world over, and its price tends to increase at a higher rate than other commodities. There is thus a powerful incentive, as we have seen in Chapter 6, to reduce the ships' companies of modern warships. Modern vessels may be required to prevent units being laid up for lack of crews. Navies have to run even to stand still.

Although the quantities of resources allocated to naval defence by the states of the world will not increase as quickly as their naval staffs would like, it seems unlikely that there will be a substantial dismantling of naval capabilities in a world that will still rely to a considerable, and quite possibly increasing, extent on the various uses of the sea. 'Getting the best value for money . . . to be had' will remain a fundamental of naval policy making. In this context Pugh argues that, 'continuation of present trends will eventually force all but the superpowers into essentially regional capabilities

Economic stimuli and constraints

and, finally, only the economically strongest of these will be left as the one naval power with global reach. However, ... these developments will, at present rates, take many decades and almost a century, respectively.' However, he also concedes that 'both political and economic circumstances might change before then'.[31] They may in fact be changing now. As we saw in Chapter 7, the economies of the superpowers are declining relatively to other states. As we have seen, the most important naval power in the world is considering swingeing cuts in capability. As the superpowers decline in relative terms the lesser states will inevitably rise in relation to the big two. This may well make it easier for them to deploy power at a distance. This, again, as we said at the outset of this section, is the economic foundation of the future promised multipolarity. Sea power is not about to price itself out of the market, indeed it looks like becoming more widely dispersed. As Pugh concluded, 'a broad spectrum of navies will long continue to exist with each being of a size and strength commensurate with the ability of their national economies to pay for them'.[32] These navies will be of considerable use to their owners in both peace and, if deterrence at various levels fails, war.

Part V
Navies in peace and war

Chapter 9

Navies in peace

Although navies are built primarily for war they find their main utility in peace, in deterring the outbreak of major and minor conflicts, in exerting influence and pressure as part of normal 'peaceful' diplomatic activities, and in enforcing the remarkable developed corpus of international law and regulation that prevents the maritime environment dissolving into chaos. Indeed, it might be argued that the world's navies are nowadays primarily configured for peace rather than war. Ballistic missile submarines pose formidable deterrent threats that if carried out could bring nothing but disaster on their owners. Even the largest navy in the world, that of the United States, concentrates its resources on forces that make more sense in a 'power projection' role in peace or limited war than in the less likely conditions of a major conventional war with the Soviet Union. The new American Maritime Strategy is an attempt to deal with the conundrum of how to use such a peacetime navy to achieve the immediate wartime objectives of safeguarding sea use.

If there is a tension between peacetime and wartime requirements in the configuration of military navies, the differing priorities within the peacetime roles themselves create extra complexities still for states deciding the type of maritime capability they ought to deploy. There can be little doubt that the constabulary and regulatory naval roles will grow steadily in importance as the forms of sea use become more diversified, and as states become more aware of their rights and duties in the various offshore zones. This diversity of naval constabulary duties, defined by Richard Hill as information gathering, the giving of instructions and orders, inspection, detention and disaster control, requires a sufficient set of skills for it to be considered to be a specialized task in its own right.[1] The United States, as is well known, has a totally separate naval organization for this role, the US Coast Guard, and this pattern has been followed by other countries, e.g. Argentina, Canada, Japan, Taiwan, and India. The creation of the 200 mile Exclusive Economic Zone acted as a potent stimulus for the creation of such specialized forces in the 1970s. Very small countries, e.g. most Caribbean states, only bother with the maintenance of a 'coast guard' or self-consciously styled maritime police force rather than a fully-fledged 'military' navy. Even those countries which tend to give the 'navy' the main constabulary role may have small additional 'navies' of various types, e.g. in the United Kingdom the Department of Agriculture and Fisheries for Scotland operates a small flotilla of ships and aircraft on fishery protection duties, and there are almost a dozen small craft operated by the Customs and Excise Marine Division. Other non-official organizations, e.g. The Royal National Lifeboat Institution, have an important part to play in maritime rescue. Although small specialist organizations may have advantages in concentrating on one task and doing it very well, too many groups can create problems of co-ordination and control.

The particular bureaucratic framework of maritime law and order will probably remain a function of local tradition and political culture, as well as the resources available. Probably greater co-ordination will be necessary in those states that do not have 'umbrella' coastguard or 'maritime safety' agencies. It is likely, however, that the competition for resources in most states between the 'constabulary' role of navies and the more 'military' role will grow more intense. The characteristics of the police vessel or aircraft are not the same as those of equipment optimized for higher level combat. Navies are increasingly going to be caught on the horns of a dilemma. Hiving off less attractive naval roles to other services might well mean a loss of resources in the short term or, as time goes by, a loss of national relevance. If the navy retains a comprehensive set of constabulary duties, however, it may end up, like the Mexican Navy for example, looking and acting like a coastguard anyway.

Even if a separate coastguard is set up the line of demarcation may not be clear. Canada provides an interesting example of this, a case study with particular piquancy in that the main ally in the military area is a major adversary in the other. The Canadian Navy is primarily oriented to contributing ASW surface escort groups to American sea control forces in both the Atlantic and Pacific (but largely the former). Given the relatively low levels of defence provision in Canada in recent years, this has not been an easy task but currently, despite design delays, the Canadians have an impressive programme of naval modernization under way. In 1985, however, the Americans sailed a Coast Guard icebreaker through the Northwest Passage on a 'freedom of the seas' operation, similar in intention, though somewhat less dramatic, than those carried out about the same time in Libyan waters. This led to such a public outcry that the Canadian Prime Minister had to announce planning for greater Canadian 'naval activity' in the Arctic to safeguard national sovereignty. The eventual proposal to purchase nuclear-powered submarines was a direct result of these sensitivities as the threat to Canadian sovereignty had subsurface and military overtones, but there have been controversies over the interface between constabulary and military roles, especially over the ownership, equipment, and role of the large, new *Polar 8* icebreaker, the most powerful in the world, currently earmarked for the Coast Guard. It was clear that the Canadian Armed Forces had serious reservations about being forced to take this ship with its significant manpower requirement and repertoire of 'civilian' roles.[2]

As the use of the sea grows, even in inhospitable northern and southern latitudes, this kind of choice will become more and more common. Navies may well have unwelcome and unfamiliar roles thrust upon them. To an extent this has already happened at the other end of the scale of peacetime naval roles, the provision of ballistic missile submarines for strategic deterrence. Perhaps only in the Soviet Union, where alternative naval traditions were weak, was this role wholeheartedly welcomed by the naval leadership. One can detect even in the United States a distinct prejudice against the SSBN role, marked by its relative low priority in the Lehman naval building programmes, and the development of alternative forms of nuclear delivery, cruise missiles, more easily integrated with the navy 'proper'. In Britain, where the potential opportunity costs were even greater, the anti-SSBN feeling is commensurately deeper. There was comparatively little naval desire to acquire Polaris, and there have been distinct signs of anxiety over the effects of the Trident programme on more 'traditional' general purpose naval capabilities. There have been persistent demands for some kind of separate funding arrangements for the SSBN force, as was achieved in the 1960s and 1970s.[3]

That admirals and their supporters should argue this is as understandable as that their political and financial masters should argue the opposite. For SSBNs are some of the largest and most imposing assets a modern navy can deploy. Their existence reflects two fundamental facts about the modern situation: first, that war between the major powers is so destructive as to be best avoided, and, second, that in an age of continental superpowers sea power can only affect the core interests of the major actors by being the least vulnerable platform for the catastrophic application of air power against their heartlands. They are thus the most important part of the 'navy' in the eyes of political leaders, and perhaps deserve to be taken a little more to the hearts of their operating institutions.

There are, however, major disadvantages to deploying ballistic missile firing submarines. The opportunity costs in other naval capabilities are usually significant and the special nature of the SSBN makes it useless for other roles. The majority of officers in the operating navies, therefore, are primarily concerned with other ways of exerting force at sea, and usually put a higher priority on these than on a high quality strategic nuclear deterrent. It is arguable whether countries with close alliances to superpowers need deploy 'nuclear deterrents' at all, or if they do whether they need to be much more than a few nuclear-tipped cruise missiles mounted in submarines that would be forward deployed anyway and which could act in desperate circumstances as a trigger for Armageddon. It would be very difficult for a Soviet leadership to distinguish a 200 kiloton nuclear burst over a major Soviet fleet base, for example, as coming from a US or a British submarine. Such nuclear forces would also be useful in deterring potential Third World nuclear powers, a factor raised at times by some supporters of medium power SSBN forces.

All nations with navies are much more interested in threats and applications of force at a much lower level of intensity. In some cases the connection between the use of force at the lower level to much higher potential levels is taken for granted. In others the dangers of escalation are much less. It is in the latter cases that violence will more probably be used, although the international laws of self-defence and proportionality limit too easy a recourse to armed violence (if not violent physical contact between ships).

Certain navies and coast guards, like the Soviet and Argentine, have shown a disconcerting tendency to ignore such rules in enforcing their sovereign and/or economic rights in off-shore zones subject to their claims. It is hard to gauge whether this will become more the international custom or not. The degree of restraint shown in the Anglo-Icelandic contenders during their fisheries dispute was remarkable given the bitterness and intensity of the conflict. Possibly, however, as lines of demarcation become more marked at sea and the legitimacy of national claims over it become more accepted, the confidence shown by the 'maintainers of law and order' on it will increase. The degree to which this will include easy resource to arms will then depend on the national culture of the state involved: this could mean more shooting rather than less.

It is not only rogue fishermen and smugglers of various kinds who are potential targets for this hostile fire. Ships are still subject to violent robbery, especially in areas where 'trading patterns bring considerable numbers of merchant ships into contact with the developing world where there are large numbers of people still living at low subsistence level'.[4] Many of these piratical raids occur in territorial waters where there is currently neither the capability, nor sometimes the will, to take serious counteraction. There can be little doubt that the growing divorce of mercantile shipping from military shipping has been a prime factor behind this violence, which has shown some disturbing trends in recent years. Established naval powers are nowadays rightly reluctant to act in waters claimed by other nations. When a local actor finally desires to take a strong line, however, the effect can be dramatic. Nigeria was able by combined naval and police action to reduce to negligible proportions by 1984 the number of piratical acts which had been running at eight a month in the worst period of the preceding years. Given the likelihood of ships remaining the interface between the developed and developing worlds and the inherent vulnerability of vessels alone at sea to armed attack, a constant effort will be required by states to fulfil their maritime law and order responsibilities. Sometimes, in situations of crisis, as with the massacre of the Vietnamese 'boat people' in the 1980s, firm international action may be required.[5]

'Piracy' proper is the use of violence for private gain. When the aim is political it gains legitimacy, although not to those against whom it is directed who tend to stigmatize its lower level acts as 'terrorism'. Happily, there has been more fear of 'terrorist' activity at sea, e.g. against oil rigs, than actual 'outrages'. Guerilla warfare at sea requires rather more infrastructure than that ashore. Ships have been hijacked, notably the *Achille Lauro*, but, as Professor Wilkinson has pointed out:

> There are a number of reasons why ship hijacking has not caught on. It is far easier to attack targets ashore, while the impact on the mass media is far greater on land. . . . For those who seek more spectacular targets, there are still enough loopholes in airport security to make aircraft hijacking possible. If aviation becomes more difficult, terrorists will be driven to other forms of hostage taking on land. Most terrorists are landlubbers and have no taste for the storms and other rigours of the high seas. There are two other key tactical considerations that would deter the average small terrorist group. It is very difficult for a tiny group of hijackers to take and keep control of a major ship carrying a large crew and perhaps hundreds of passengers . . . and it is in principle far easier for the authorities to mount a counter attack and regain control of the vessel.[6]

The more organized guerilla groups have attempted maritime activities, notably the PLO, which has attempted raids with vessels from freighter size downwards, but it usually requires state backing to acquire the resources to engage in maritime warfare, as when the Libyan freighter *Ghat* unloaded mines in the Red Sea on behalf of the Islamic Jihad in 1983. Some eighteen ships were damaged.

The international naval operation designed to neutralize these weapons is but one example of the use of naval forces in counter-terrorist and other low-level operations. The growth of the Israeli Navy has received considerable encouragement from the maritime PLO threat and the alternative to bombing that naval power provides for relaliatory actions. The US Navy has made something of a name for itself in the 1980s as a 'counter-terrorist' force. When US forces were subject to 'terrorist' attack in their 'peacekeeping' mission in Beirut, the battleship *New Jersey* was called upon to wreak vengeance with her 16 inch guns. The use of the spectacular battleship had the effect of demonstrating to public opinion back home that 'something was being done'. There was also reduced danger of suffering losses, a problem demonstrated when naval aircraft suffered losses while engaged in similar missions against the Syrians. The actual effect of the dubiously accurate bombardment was less certain, and probably less important.

The bombing of Libya by a combined USAF and carrier air strike in 1986 was another example of naval forces being used for 'counter-terrorist' bombardment. More discriminating was the use of naval interceptors to capture the *Achille Lauro* hijackers over the Mediterranean.[7] Generally, if the main target is the actual 'terrorist' rather than the opinion back home, it is the less spectacular riposte that is the most effective deterrent. One suspects that the training of maritime anti-terrorist squads able to deal directly with active maritime terrorism acts as a greater deterrent than the possible attacks of an American carrier battle group or battleship on one's backers. With the drawing up in 1988 of international conventions to prevent terrorist attacks on ships on the high seas and fixed maritime platforms, there might even be some prospect of joint international action to deal with threats that are increasingly being perceived as universal and not part of legitimate great power competition.

Naval bombardment of weaker states that cannot hit back effectively is a classic form of gunboat diplomacy. The latter, as we have already discussed in Chapter 7, shows little sign of going away. It used to be fashionable to argue that its days were numbered with the spread of relatively well organized sovereign states and the acquisition by them of increasingly powerful defensive weapons. Yet the record of the 1980s clearly demonstrates that if the margin of technological superiority is sufficiently great, a first class navy should have little difficulty in 'punishing' even quite a superficially

Navies in peace

strong littoral state. The Iranian Navy was consistently overwhelmed in surface action by American ships and aircraft. Events off Libya show clearly that an American carrier battle group has relatively little to fear from a few isolated missile boats or badly handled aircraft. It can operate almost at will.[8] It will take some considerable time for the margin of superiority of the US Navy at least to be eroded sufficiently for it not to have considerable scope for activities ranging from simple bombardments to amphibious landings like the invasion of Grenada in October 1983. Indeed, one of the primary roles of the US Marine Corps remains, and is likely to remain, the maintenance of US interests, by force if necessary, in the traditional US sphere of influence in the Caribbean and Central America. The only weapons that have seriously discomfited the Americans have been Iran's mines, but this threat was soon overcome and the lesson of the need to improve MCM capabilities has been clearly learned, for the time being at least.

As the US Navy is currently the major exponent of naval diplomacy, and is likely to remain so for the foreseeable future, perhaps it is worthwhile to look at its own definition of the range of activities involved. They were succinctly described by Admiral Watkins in early 1986 thus: 'Maintaining presence, conducting surveillance, threatening use of force, conducting naval gunfire or air strikes, landing Marines, evacuating civilians, establishing a blockade or quarantine, and preventing interference by Soviet or other forces.'[9] Admiral Chernavin would probably only have altered one word, although the power of the Soviet Navy to engage in active action is significantly more restricted than the US Navy's. Limited carrier assets act as a potent constraint on bombardment potential. The Soviets are slowly improving their carrier air power but its potential is going to be very limited for some considerable time. The concentration of the Soviet Navy on operations against other naval forces does, however, put a very potent, perhaps potentially *the* most important constraint, on the activities of the US Navy. The most dramatic demonstration of this has probably been the Soviet anti-carrier warfare activities during the 1973 Arab-Israeli War, when the potential price of direct US involvement was demonstrated only too clearly.[10]

The Soviets on the whole have used their maritime forces sparingly in more positive ways in their own diplomatic interests, transporting friendly troops to reinforce client forces on land, protecting their merchant ships in danger zones, demonstrating the USSR's concern when her ships have been seized and even carrying out the odd limited bombardment in support of her friends. Nevertheless, despite the accepted wisdom of most commentators, the Soviet Navy has been less 'operational' than those of the West for maritime intervention.[11] It is significantly limited by the small size of its amphibious component. Possibly the traditional bureaucratic dominance of the ground forces has prevented the naval infantry replacing the airborne forces as the Soviet Union's main means of military intervention. When the Soviet Union does feel impelled to land forces rapidly the target may be far from the sea, e.g. Afghanistan. The USSR is thus more willing to face the higher risks inherent in airborne operations compared to seaborne landings.

As we discussed in Chapter 7, although the superpowers will remain the main practitioners of peacetime naval diplomacy in the short to medium term, other nations will also use their navies as diplomatic instruments. Both Sir James Cable and Ken Booth have produced comprehensive analyses of the peacetime uses of naval forces.[12] Booth offers three headings for the aims and objectives of the 'diplomatic role' of navies: (1) Negotiation from strength, (2) Manipulation, and (3) Prestige. The first and last are more or less self-explanatory but the second heading requires closer definition: 'changing (however incrementally) the political calculations of relevant observers. . . . Effects may be induced by negative as well as positive actions'. Naval forces might be disposed to demonstrate support to friends, 'gain or increase access to new countries', create 'a degree of naval dependency', or establish the 'right to be interested' in a distant area.[13] All of the above imply no actual action being taken. Booth lumps all *uses* of force under the 'military role'.

Cable puts the distinction in a different place. He defines 'gunboat diplomacy' as 'the use or threat of limited naval force other than as an act of war'. The distinction from 'showing the flag' is clearly made in a recent article:

> The expression 'showing the flag' is often loosely employed as a synonym for gunboat diplomacy. That practice involves the use or threat of limited naval force for some specific purpose. Even the 'expressive' variety, when warships are manoeuvred merely to ventilate emotion, is recognisably related to a particular dispute. 'Showing the flag' is a more general reminder to foreigners of the existence of the navy concerned. Any threat of force is seldom more than remotely explicit.[14]

To Cable, therefore, the existence of the *specific* dispute is the central defining factor, both distinguishing between gunboat diplomacy and generalized flag showing at the one extreme and specific uses of limited force, and 'war' at the other. Cable defines the distinction between gunboat diplomacy and war as follows:

An act of coercive diplomacy ... is intended to obtain some specific advantage from another state and forfeits its diplomatic character if it either contemplates the infliction of injury unrelated to obtaining the advantage or results in the victim attempting the infliction of injury after the original objective has been either achieved or abandoned. Coercive diplomacy is thus an alternative to war and, if it leads to war, we must not only hold that it has failed: we may even doubt whether it ever deserved the name.[15]

In 1982 General Galtieri tried coercive diplomacy: the British replied with a limited war.

Booth's distinction seems to centre around means, Cable's around ends, although the latter does slightly muddy the definitial waters when he argues additionally that 'showing the flag becomes gunboat diplomacy when the risk of encountering armed resistance is deliberately run'.[16] The implication must be one only runs the risk of resistance when one has a strong specific motive for doing so.

Cable divided his 'gunboat diplomacy' into four kinds:

1. *Definitive* – when a *fait accompli* is created;
2. *Purposeful* – to induce 'someone else to take a decision which would otherwise not have been taken: to do something or to stop doing it or to refrain from a contemplated course of action';[17]
3. *Catalytic* – 'a situation arises pregnant with a formless menace or offering obscure opportunities. Something it is felt is going to happen, which might otherwise be prevented if force were available at the critical point. Advantages, their nature and manner of their achievements still undetermined, might be reaped by those able to put immediate power behind their sickle. These are situations peculiarly favourable to the exercise of limited naval force';[18]
4. *Expressive* – 'to express attitudes, to lend verisimilitude to otherwise unconvincing statements or to provide outlets for emotion'.[19]

Cable then goes on to draw up an ascending list of levels of operation: simple ship, superior ship, simple fleet, superior fleet, simple amphibious, and superior amphibious.[20] As one moves up the levels of violence the numbers of fleets able to carry out the required operations diminishes very significantly, at least when one moves over any distance. When one moves even to 'simple fleet' operations the effort can be considerable, e.g. at the time of writing a navy like Britain's has to tie up a third of its surface combatants in maintaining a simple fleet in the Gulf. Both numbers of ships and sufficient auxiliaries are required to achieve the sufficient 'reach', at least in national terms.[21]

The dividing line between 'simple fleet' and 'superior fleet' operations is a function of the opposition to be expected and it is here that the growth of naval capabilities among potential victims acts as an important factor. Deploying a 'superior fleet' would strain even powers like Britain in many sets of circumstances. Simple (i.e. unopposed) amphibious operations are also by definition dependent on the nature of the opposition but may be carried out with surprisingly limited forces (e.g. a single ship's contingent of marines) in the right circumstances. Superior amphibious operations, however, require maximum efforts from medium powers, and are correspondingly risky.

The growth of navies with significant abilities to project power, and the increasing opportunities the future multipolar world, may provide for this power to be projected, will widen the number of assailants in exercises of gunboat diplomacy and the types of operations that are practical for them as well as the number of victims. India, China and perhaps even Japan may find themselves as practitioners of the art, as well as other local powers where appropriate. The Europeans will also retain a surprising number of options if they choose to operate them, especially if they pool their efforts 'out of area' as they do within it.

Before leaving the peacetime roles of navies we need to examine the roles they might play in the management of a major crisis between the superpowers and their allies. The American Maritime Strategy puts great emphasis on crisis response and the world-wide deterrent value of forward deployed forces. Admiral Watkins argued that, 'the flexibility and precision available in employing naval forces provides escalation control in any crisis, but has particular significance in those crises which involve the Soviet Union'.[22]

NATO's Concept of Maritime Operations, as well as plans to fight (see below) also contains a 'family of contingency plans for use in times of tension'.[23] These are, according to official spokesmen, 'extremely flexible in the scale of forces they may generate, and how and where they may be applied. They permit graduated responses to a threat, providing a wide range of political as well as military options, with the aim of deterring war.'[24] Once the NATO Defence Planning Committee approves, SACLANT can assemble forces from Alliance members for early deployment. Such forces might well include short notice amphibious units (see Chapter 10). Surveillance, however, would be crucial. In the words of Deputy SACLANT in 1987:

> The presence of allied forces in a surveillance role not only allows the collection of information on Warsaw Pact platforms, weapons

systems, operational patterns, tactics and current activities; but also provides an efficient and effective demonstration of our own maritime capabilities. The advantage of these contingency plans is that they are flexible and are written to move easily into either complementary war plans or, preferably, back to peace.[25]

This process of maritime crisis management might be significantly enhanced by the increasing ritualization of crisis responses through an agreed regime of confidence and security building measures (CSBMs) putting agreed limits on naval exercises in normal times and providing codes of conduct for naval manoeuvres in close proximity. As Admiral Mustin has perceptively pointed out, this process has tacitly gone a long way already with the routine deployment on exercise of US carriers in the sea areas relatively close to Soviet bases, in a sense legitimizing their forward deployment for deterrence and stabilizing crisis management purposes when circumstances require.[26]

Formalizing these deployments in a regime of CSBMs might enhance stability still further. It would provide formal legitimacy to forward deployments, establishing their 'normal times' nature. The onset of a 'crisis' could also be formally announced by exceeding agreed limits or, at least, manipulating the threat to do so. Once 'normal times' limits were exceeded, strict rules of engagement would be even more important to prevent the situation getting out of hand. This might include assets being exposed to risks that would not be acceptable in a normal operational situation. There is little reason to think that the political imperative not to be the first to fire would be any weaker in such a confrontation than it has been in other post-war crisis situations, e.g. in the Gulf where the USS *Stark* was a notable victim of necessary restraint. Indeed, with the stakes so high restraint would be even more important – and strongly applied.

In Europe, however, persistent and contradictory doubts have been expressed not only about the wisdom in crisis stability terms of US forces rapidly deploying forward in a crisis, but also whether they would actually be available to do so given the other likely calls upon them. This gives Europe's own navies a vital role in early forward crisis deployment in the Norwegian Sea. Such deployment demonstrates collective Alliance resolve by its very presence (and the insertion of amphibious reinforcements), but it does not create such a dangerous offensive threat that the adversary in the crisis need feel provoked. The possibility of US reinforcement would, however, be real; indeed, the NATO ships would be actively engaged in 'precursor operations' to clear the way for carrier striking fleet. One would hope that the Soviet Union would take the hint that they had misjudged the situation and defuse the confrontation. Such operations are much less demanding than a full-scale battle in the Soviet Union's 'backyard', but the naval forces available and the land-based air cover giving them support need to be sufficiently strong not to pose tempting vulnerabilities. Such a level of capability is not too difficult to achieve with present and likely European naval assets.

Despite everyone's efforts, things might go wrong and shooting might start. A victim of gunboat diplomacy might surprisingly refuse to accept defeat. An assailant might grossly misjudge the situation. Despite the ritualized manoeuvres of warships – or perhaps even because of them – a quarrel totally unconnected with the sea might spill over into a situation recognizably called 'war'. What of navies then?

Chapter 10

Navies in war

Navies would have a part to play in all forms of future war, total and limited, nuclear and conventional. To take the least likely first, if an unlimited nuclear war occurred naval forces would be the ultimate instruments of the destruction of the participants' homelands. The ballistic missiles carried in the submarines' missile tubes of both superpowers are enough on their own to obliterate the others' societies as viable units. Although the United States has currently rather neglected its SSBN forces in its rush to enhance general purpose naval capabilities, the end of the 1990s should see forces that could be used, if required, in counter-force strategic nuclear warfare against land-based missiles, Soviet command centres and other 'hard' targets.

This might enhance the 'strategic reserve' role of the seven hundred plus nuclear-tipped cruise missiles to be deployed in the US Navy's attack submarines, battleships, cruisers, and destroyers. Such missiles provide a capability to strike the Soviet Union from all azimuths, compounding the defensive problem to impossible proportions. The numbers and flight characteristics of cruise missiles make them relatively invulnerable to a wide range of Soviet weapons, especially the expanded ABM system that the USSR might deploy in any break out from, or breakdown of, the ABM treaty. There are significant costs, however. Naval forces are likely to be called away from more immediately useful duties to pose a nuclear threat; if they are not specially allocated to nuclear strike duties they are unlikely to be in the right place to attack the required land targets. Movements of nuclear-armed surface units might well be constrained, either out of prudence in a crisis or because they are unwelcome in nuclear free areas. Nevertheless, the attractions of using the West's maritime predominance to enhance the strategic nuclear threat facing the Soviet Union seem at the moment to be too strong to overcome.

The Soviet Union continues to improve its nuclear-powered

ballistic missile submarine forces with new missiles, MIRVs and boats (if boat is an appropriate term for the huge Typhoons). These forces seem to be regarded by the Soviets as a strategic reserve to be withheld and defended in bastions off the Soviet coast and below the ice pack. Deployed in a more forward position, perhaps implying earlier potential use, will be the large and small strategic nuclear cruise missiles currently about to enter service in reply to US developments. Soviet sea-launched cruise missiles might join Soviet land-based forces in a counter-force first strike against the USA should the USSR's leaders be forced to such disastrous extremes.

The smaller nuclear powers, too, stand ready to inflict catastrophic damage on their opponents if suitably provoked. Currently constrained to aim all its available Polaris warheads at Moscow, the British will be able to adopt more ambitious targeting plans when Trident D-5 comes into service. The French will soon be able undoubtedly to destroy the *œuvres vives* (vital works) of their Soviet potential opponent with their MIRVed M-4 missile and its eventual M-5 replacement.[1] China is also sending some of its strategic nuclear forces to sea with an SSBN in service and three more are under construction, but the programme has run into considerable and persistent difficulties and it proceeds very slowly. All these smaller powers would no doubt continue to target their weapons primarily on relatively soft, urban industrial targets so as to inflict maximum damage. More importantly, however, the *threat* of the use of all these ultimate deterrents would be used as the primary means of deterring all forms of war, or should war nevertheless occur, limiting it to 'acceptable' levels, whether that be very limited nuclear exchanges at a strategic or theatre level or, much more likely, to conventional weapons alone.

Navies have an enormous interest in such limitations as they are at their most relevant in *conventional* hostilities. Both superpowers, to a considerable degree, and the British and French to a much more limited one, deploy tactical nuclear weapons for naval warfare. It is possible to rehearse a theology of nuclear war at sea that has some plausibility (lack of collateral damage, easy definition of military targets, etc.), but in reality it seems unlikely that nuclear weapons would be released for use at sea before they were released as the ultimate symbol of resolve on land, probably in Europe or Asia. Given the prevailing doctrines of nuclear use whose logic is more in line with the subtle French concepts of 'pre-strategic' nuclear warnings rather than simple operational nuclear defence, it would make sense not to blur the signal too much by, for example, using nuclear weapons in an early stage of hostilities as ASW devices. There might, however, be some utility in making nuclear warnings against naval targets if these were of sufficient significance. The sinking of a large surface warship like *Kirov* or *Kiev* with a large nuclear warhead, such as the 300 kiloton weapon carried by the French ASMP missile (which is to be fitted to naval aircraft), might well be a sufficiently 'sensational' event to give the Soviets pause for thought.

If such a warning failed and full-scale nuclear war ensued, a major target of counter-force attack would be the ballistic missile firing submarines of both sides. Given the inherent capacity of such submarines to lose themselves in the ocean, only isolated successes could be expected even if nuclear weapons were used (including long-range strategic missiles in barrage mode if such weapons did not have more appropriate targets or objectives of higher priority). There are, and there are likely to remain, too many variables operating in favour of the ballistic missile submarine to undermine the second strike potential of even quite a small force of these boats. They have the advantage of not having to close their targets to be effective. This makes the ASW problem enormously difficult. A surprise disarming strike is also probably impossible. Once an SSBN has fired its first missile it has given away its position, but in full-scale nuclear war, with all its confusions, it is asking a great deal for an enemy to react rapidly to ballistic missiles rising from the ocean in a number of widely separated places. An individual submarine, therefore, will be likely to get away all its missiles. This, however, implies Herman Kahn's nuclear 'wargasm'. Submarine launched ballistic missiles are probably not the most appropriate means of mounting nuclear demonstrations. Maintaining the uncertainty of their positions is paramount and their owners would probably always possess alternative means of carrying out nuclear warnings.

It is possible that some kind of naval hostilities might continue after the destruction of the homelands of the participants in a full-scale nuclear war. In the 1950s it was fashionable for a time, especially in naval circles, to speculate about a kind of 'broken backed' hostilities, when operations would continue at a minimal level, both to sustain survivors and to rebuild for a next intensive phase which might even bring victory. This concept was always unattractive to the policymakers or the public then, and is little less so today. It is possible that naval forces would continue to fight each other in some sort of strategic spasm until their ammunition and supplies were exhausted. Perhaps cut-off from their infrastructure, some might become latter-day pirates preying on the remaining shipping and coastal areas of what parts of the world remained habitable. The control of a coherent 'war', however, would be impossible and only sentimentality and instinct would call the ships and submarines of countries that had suffered nuclear attacks back to view the destruction.

Happily, the uncertainties and dangers of such a scenario (if the understatement is to be forgiven) make its happening extremely unlikely. Indeed, the likelihood of limited nuclear war becoming total is going to give decision-makers enormous incentives to keep hostilities below the nuclear threshold. The likelihood of large-scale conventional superpower vs. superpower coalition war is a matter of debate. Perspectives differ within NATO, for example. The West Germans, for whom such a war would bring destruction virtually indistinguishable from nuclear devastation, have always tended to argue against too much emphasis on large-scale conventional war fighting. Nevertheless, even they now seem to be accepting that the only credible response to a Soviet conventional attack is conventional response in kind. Every trend in NATO strategy is towards an enhanced conventional capability. The implications of this are clear. A conventional war will not be over in a few hours: it may last weeks or months, possibly even years. The greater the conventional emphasis in NATO strategy, the greater the importance of sea power in its traditional role of maintaining the flow of men and supplies so that the flight can be continued and domestic populations sustained.

It can be argued that a long drawn out conventional war is not a very attractive option, and that it should be prevented by making clear that nuclear escalation is almost certain if the conflict goes on beyond a certain relatively brief period of time. In an era of nuclear parity, however, such a threat is only dubiously credible. There is probably no alternative to retaining at least some options on sustaining hostilities for as long as possible. Deterrence is indeed enhanced by demonstrating to the potential enemy, especially one with a plausible, short conventional war winning operational doctrine, that there is a more likely danger than nuclear war if he chooses to try such an adventure: a long drawn out conventional conflict which his economic weaknesses mean he cannot win.

The self-cancelling effect of both side's nuclear forces is making large-scale conventional war at sea more, rather than less, likely in the event of superpower confrontation: although that confrontation itself is something both sides would rather avoid, due to the incalculable dangers of nuclear escalation in the event of a direct clash between them. Even as things stand today, with publicly articulated statements emphasizing the high probability of nuclear escalation, NATO plans to carry on a major maritime reinforcement operation of conventional forces. The current Rapid Reinforcement Plan calls for thirty US and Canadian combat brigades, 100 air squadrons, and 300,000 men to be transported to Europe after the day reinforcement begins (R-Day). By R + 180 some 1½ million men, 8½ million tons of ammunition and stores, and 114 million barrels of fuel would be transported to Europe. In addition to this about 100,000 military personnel, 25,000 tons of material, and perhaps 20,000 vehicles would have to be transported from the United Kingdom to Europe. Ninety-eight per cent of personnel can be airlifted but, even with pre-positioning of significant amounts of heavy equipment in Europe, very large quantities of the latter must come by sea. Shortages of aircraft and congestion problems at airfields put significant constraints on reinforcement airlift.[2]

According to EASTLANT (Eastern Atlantic) figures, the requirement for military reinforcement and supply shipping over the first six months is very considerable: 3,045 shiploads.[3] The pressure for ships would be highest in the opening stages. In February 1988 a British MOD official estimated the first month's shipping movements at 1,000 and cargoes at 1.3 million 'tonnes of war stores'.[4] SACLANT's (Supreme Allied Commander Atlantic) figures are rather smaller but this could reflect recent changes in perceived reinforcement requirements. At the 1986 Sealink Conference SACLANT, Admiral Lee Baggett stated that 'our latest studies show that the requirement is about 400 shiploads a month for military reinforcement and 400 for military re-supply: according to 1987 figures these 800 ships would carry 40–50 million tons of cargo'.[5] On given shipload figures even this pace would only be necessary for the first two months. After R + 60 one might, therefore, expect the military requirement to tail off somewhat, perhaps to about 300 shiploads a month or so.

In addition to all this, economic shipping would have to continue to arrive to maintain economic life in embattled NATO. SACLANT currently estimates the total number of monthly shiploads that need to arrive in NATO's ports at between 1,500 and 2,000.[6] Unofficial estimates of NATO's 'civilian' requirements in war are 365 million tonnes per annum of crude oil, bauxite, iron ore, phosphates, cement, sugar, salt, fertilizer, gypsum, non-ferrous ores, and wood products for North America, and 948 million tonnes per year of similar materials plus coal and grain for NATO Europe. More official SACLANT figures estimate the economic requirement at '95–110 million tons of cargo'. As Vice-Admiral Sir Geoffrey Dalton, Deputy SACLANT, told his audience in Oslo in April 1987:

> For the economic shipping . . . the routes will be quite different from previous wars. No longer will the great majority of cargoes move from North America to Europe, instead they will originate

mostly in other parts of the world. The reason is that, although large quantities of items like, grain, coal, iron ore and general cargo will continue to be transported to Europe from North America, by far the bulk of the economic shipping will travel to Europe from the Arabian Gulf, West, North and South Africa, South America and Australia. Moreover, as an additional difference from previous experience, the destination of a significant part of the shipping will be North America.[7]

The shipping requirements implied by the above needs are hard to calculate exactly given different ship sizes, speeds, and transit times. The NATO reinforcement and resupply requirement is officially put at 600 American and 600 European merchantmen, but there are doubts over whether these figures can now be met.[8] According to the US Joint Chiefs of Staff's own figures, the total number of 'funded strategic sealift resources' available is 533 ships. Reserves requiring two to three months to get to sea and, hence, only useful for replacing attrition, add another 197 hulls. These must cover all contingencies, both in the Atlantic and Pacific. Korea can add another 45 ships, and 600 vessels are expected from European NATO.[9]

Sadly, however, the latest figures produced by NATO's Planning Board for Ocean Shipping show that only 448 of the requisite dry cargo ships are available on the NATO 'Sealift Shipping List'.[10] The situation is not quite so desperate as it sounds. The 600 figure included a 50 per cent margin but that margin is now only just over 10 per cent and falling: NATO lost some 50 dry cargo ships in 1986–7 alone. It is now being recommended that beneficially owned flag of convenience vessels be added back to the list in order to restore its size. This is especially important, as the sealift requirement is expected to increase.[11] Whether or not the ships are available, however, their defence plus the defence of all the economic shipping required for the West's survival raises all the old questions of battlefleet versus direct protection as discussed in Part I.

This debate has been sharpened in the 1980s by the American development of the forward Maritime Strategy, its clear articulation in open publications and its adoption as the basis of NATO's maritime defensive concept.[12] Although, as we have seen, World War II was a triumphant vindication of convoy and escort, the predominance of this operational doctrine was soon questioned. The major factors were technological developments both in submarines and anti-submarine warfare, the advent of nuclear weapons, and the bureaucratic domination of the aircraft carrier in the dominant navy of the Western Alliance. New types of high-speed submarines, both conventional and nuclear powered, rapidly undermined many of the tactics of the World War II convoy escort. The development of new long-range surveillance sonar networks seemed to provide one answer to this, giving ASW some of the characteristics of air defence in allowing the vectoring of 'interceptor' assets on to fast-moving threats. Hence the 1950s saw a major new emphasis on 'hunter killer' operations. Even these, however, seemed less effective than the 'attack at source' on the submarines in their bases. These operations had been a miserable failure in both world wars but in the 1950s nuclear weapons, including special naval earth penetrating devices developed for the role, accurately delivered by carrier aviation, could open up the most formidable submarine pen. As soon as NATO's Atlantic Command was set up, a nuclear-armed carrier-striking fleet was established as its primary arm, with the 'attack at source' on the Soviet submarine fleet as a high priority. This fitted in well with an understandable tendency of the USN to emphasize the carrier battle group operations with which it won the Pacific War, rather than the convoy escort work that it had largely left to its British allies.[13]

There is thus little really new in the US Navy's Maritime Strategy and its proposals for forward carrier offensive, although that offensive is now designed to be primarily conventional rather than nuclear. NATO has always had a carrier-centred battle fleet with an important perceived role in gaining the command of the North Atlantic in any future war. With the new-found confidence of the 1980s, however, the US Navy now openly discusses how such an offensive would force the mass of Soviet naval assets on to the defensive, so drawing them away from NATO's Atlantic shipping. Not only would the Soviet Navy be pinned down, but it might be rapidly defeated and destroyed in a latter day version of Mahan's decisive battle.

It had tended to be assumed in the 1960s and 1970s that carrier battle groups (CVBGs) would keep their distance from heavy concentrations of enemy submarines and strike aircraft due to the inherent vulnerability of the carriers and their escorts to torpedo and missile attack. The authors of the Maritime Strategy have turned this on its head. They now argue that a CVBG would positively welcome the enemy coming out to fight with his major offensive assets (effectively a three dimensional battle fleet) as the technologically superior NATO forces would use this opportunity to destroy them. The Battle of the Philippine Sea, when US carriers effectively destroyed the Japanese carrier-based Naval Air Arm as an effective force, is a powerful precedent. The long-range threat posed to Soviet bases by the carrier-based aircraft (and to a lesser extent cruise missiles) of the latter-day American battlefleet would mean that the option of mounting a 'fleet in being' might not be so open to the Soviets as it was to the Germans in both wars. They would have to come out and fight or face destruction in harbour. The strategic 'pay off' from the naval offensive might be greater still. What *is* new in recent formulations of US naval strategy is the focus on the Soviet ballistic missile firing submarines. Some American naval planners seem to have detected in missile firing submarines a key form of Soviet sea use which the Soviets would not only hazard their fleet to defend, but which might even be a point of such sensitivity that it could be exploited to achieve war termination on favourable terms, i.e. victory. This campaign could lead well under the polar ice cap. The Soviets have understandably chosen to exploit one of the advantages geography has given them of a large area of ocean cut off from aerial or space surveillance by solid masses of ice. Modern Soviet PLARBs are specially configured for under ice operations, and the Americans have taken up the challenge by modifying existing and future SSN designs to go in after them. The cold arctic depths may well be a major combat theatre in any future conflict.

The 'war termination leverage' achieved by inflicting attrition on Soviet ballistic missile firing submarines is one of the most controversial aspects of Maritime Strategy.[14] Some, like Barry Posen, have seen it as a formula for premature and accidental nuclear escalation.[15] The Soviets might see the loss of their SSBNs (in Russian, PLARBs) as tantamount to a pre-emptive strategic strike. Alternatively, it can be argued that the Soviets themselves expect to destroy NATO nuclear targets on land in their official conventional offensive without necessarily leading to NATO nuclear use. There would apparently be no automatic 'use 'em or lose 'em' assumptions in the minds of the USSR's strategic planners.

A major offensive into waters close to the USSR is fraught with dangers and the US Navy has been at pains to point out that it foresees no simple 'charge of the light brigade' with its carriers. Examination of recent NATO exercises clearly demonstrates that in the North Atlantic, at least, the aim is to take the initiative early and place carriers inside the fjords north of Bodo, where the mountains and water conditions help the NATO forces exploit the traditional advantages of the defender. Soviet attack options are constrained by geography, and available NATO assets can be concentrated on certain threat axes; Soviet attempts to neutralize the potent threat these forces pose to Kola bases are thus more easily defeated. Once the Soviet three-dimensional battlefleet has been neutralized then moves forward might be possible.

This brings in a further danger, however. If the forward deployed NATO Striking Fleet did succeed in achieving a Marianas Turkey

Shoot against the Northern Fleet's aircraft and inflicting serious casualties on submarines and surface ships, it might only leave Soviet commanders with nuclear options with which to defend the homeland and their vital nuclear strategic reserve. Soviet political leaders, faced with dire and extreme choices, might well feel that a definitive nuclear strike against the NATO Striking Fleet might not necessarily lead to all-out nuclear war, although this assumption would be more difficult to make with the fleet operating close to land. The American reaction to the loss of tens of thousands of naval personnel might, however, be more violent. A disastrous spiral of escalation would probably ensue.

Of course, American confidence might be misplaced. Even a conventional naval battle might go badly for the Striking Fleet in its enemy's 'backyard'. A single lucky hit from Soviet long-range homing torpedoes could have catastrophic effects on the operational viability of an American supercarriers. The level of missile threat posed by Soviet aircraft and submarines (and possibly surface ships) is such that any air defence, even the F-14/Aegis combination of the Americans, would be stretched to the uttermost, especially as it would also have to cover the British ASW forces, vital to give protection against the subsurface threat but less well-equipped with organic air defences.

Naval warfare of this intensity is a desperately risky business, with very high stakes, given the concentration of resources in each individual modern naval unit. It is likely, therefore, that prudence and caution will prevail. Western carrier battle groups are not going to go charging off into the Barents Sea. Here the initial Western offensive will be limited to nuclear-powered submarines. This will achieve the vital objective of pinning down Soviet assets, although there would still be various pulls for escalation. Submarines in these circumstances would be very hard to control if intra-war bargaining produced a cease-fire. In such a submarine offensive those lesser naval powers which have invested heavily in SSNs (at considerable opportunity cost) could, however, play a significant role.

Whatever the level of forward deployment it is unlikely that NATO could neglect 'barrier' operations to take advantage of its favourable geographical position, notably the Greenland–Iceland–UK 'gap'. All kinds of force could be used on barrier operations although mines, conventional submarines, and shore-based aircraft seem the most attractive options, leaving surface units for use elsewhere, where their unique characteristics are of greater use. The utility of barriers is, however, only limited and far from certain. It is hoped that five may be set up, but their efficiency might vary from one kill in twenty boats to one in five.[16] Moreover, barriers could, of course, only become operational once hostilities had commenced. A large number of hostile submarines could well be able to break out into the open ocean therefore.

It is often argued that the Soviet doctrine seems to indicate that they do not put a very high priority on interdicting reinforcement shipping in the allocation of their submarine and air assets. The naval forces would be tied down protecting their precious PLARBs in their defended 'bastions'. Certainly the Soviets seem prone to such doctrinal assumptions and this argues for NATO doing as much as it prudently can to foster such concerns in order to direct Soviet activities away from what could hurt the West the most. Nevertheless, the central importance of reinforcement and supply shipping to NATO's capacity to continue with conventional fighting should imply as high a level of capability being allocated to the direct defence of shipping as practical. Indeed, a good case exists for direct defence operations still being allocated the highest priority. Only in direct sea control operations can naval assets achieve the primary strategic aim – the safe and timely arrival of cargoes – even if they do not actually sink or shoot down enemy units. The latter's life can be made so difficult and dangerous that they cannot carry on with their missions. If it actually attacks a valuable cargo vessel an enemy submarine has inevitably to give away its position, making its destruction much more likely, especially if a friendly naval unit is nearby. Even if its initial attack is successful it will thus have considerable difficulty in carrying out a second or third. Modern ideas combining close escort and more distant support using modern passive sonar techniques have been suggested and seem very practical.[17]

One major argument against convoying, one with a very long pedigree, is shortage of escorts. It is now usual to say that NATO must adopt a forward strategy as she has insufficient escorts to protect directly all her shipping. The evidence produced to sustain this thesis, however, cannot but be very uncertain. A relatively recent study found it 'difficult at best to determine Atlantic escort level requirements more accurately than somewhere between 18 and 273'. Persian Gulf requirements were between 53 and 370 and a 'middle the road guess' for Pacific convoys 'about 175'. Non-US NATO escorts usable in an Atlantic battle were estimated at between 50 and 100. The whole question was 'a matter of judgement rather than mathematics'.[18] The current compromise seems to be, as far as the Atlantic is concerned, to send most ships individually to the Azores where they will be escorted in convoy to the coast of Europe, being defended in depth by surface and air forces.[19]

There may be little alternative but to return more wholeheartedly

Navies in war

to convoy as the declining capability of very long-range detection makes such direct protection more necessary than less. As we have seen, the long-range sonar capabilities that allowed the authors of Western naval operational concepts to downplay the disturbing lessons of experience seem to be coming to an end. Not even at the height of passive sonar capabilities were naval commanders able to do without defensive screens for their most valuable assets about which they cared most, their major warships. Long-range active sonar capabilities in helicopters, and the availability of cheap floating helicopter platforms, might well improve the effectiveness of escorts to the level that the inherent economy of the convoy system in grouping assets at risk with protective forces is still further enhanced. Even Norman Friedman, an informed analyst usually quite hostile to convoys, has argued that 'the advent of long range helicopter dipping sonar would be a way of achieving something resembling effective screening even though NATO cannot really hope to achieve ship numbers comparable with those of, say World War II'.[20] It was also interesting to see convoying immediately adopted to protect US-flagged shipping from aerial, mine, and surface threats in the Gulf in 1987.

Convoy has never been as profligate of escorts as its opponents have consistently claimed. Indeed, under the conditions of the two world wars it proved to be by far the most *economical* and *efficient* use of limited naval assets. What has often been the case, however, has been the starvation of mercantile escort groups to feed the apparently insatiable appetite of warship escort and 'main fleet' operations. Troop convoys have usually received a higher priority somewhere in between normal ships and warships; indeed, main fleet units have been used to screen sensitive troop convoys on many occasions. This would lead one to speculate that in any future conventional NATO war there might well be alternative pressures for the allocation of NATO's precious naval assets.

NATO's Concept of Maritime Operations (CONMAROPS) is in fact very flexible. It was designed as 'a conceptual basis for Nato's maritime planning in the 80s and 90s'.[21] It contains three principles:

> The first one is containment – which in war translates into offensive action to prevent Warsaw Pact forces deploying into open waters. This is particularly important in areas such as the Norwegian Sea – if we can prevent Soviet maritime forces from steaming into the Atlantic, the antisubmarine battle, particularly with our limited forces, will be less difficult.
> The second principle is defence in depth – the deployment of

forces to defend in depth is designed to put Soviet ships and submarines at risk at every turn. For this to be achieved, we must be prepared not only at the forward edges of the NATO area, but throughout all our area.

The third principle is keeping the initiative – vital in tension to ensure forces are brought early to a high state of readiness and properly positioned, but equally important in war to ensure they are used where they are needed most.[22]

This clearly implies forward operations of some kind, the readiness 'to fight the Soviets at the forward edge of the NATO area, along their exit routes, and in defence of the allied war and merchant shipping', and to adopt an aggressive operational and tactical approach. It does not, however, prescribe any particular action: available forces would be shuffled as required between three 'campaigns', Norwegian Sea, Atlantic, and Shallow Seas on the basis of a *set* of 'general defence plans'.[23] NATO's Atlantic naval exercises are ambiguous, as they are bound to be, as they are merely testing and rehearsing options. Sometimes operations in the Norwegian Sea follow strike fleet operations in close defence of 'sea lines of communication'. Sometimes the Striking Fleet heads straight for Norway.[24] The actual war deployment would depend on 'the prevailing circumstances . . . what Soviet forces are deployed in the area? What is the status of the land battle? And who has the necessary air superiority?'[25] The key to the Norwegian sea battle is early deployment. If powerful forces are not inserted early as part of NATO's crisis contingency plans (see Chapter 9) it will prove much harder to insert them against a prepared enemy. World War III may well begin, like the active part of World War II, with a race to the fjords.

In order to bring ground forces to bear in Norway, the Americans seem to envisage a three-phase approach, taking advantage of the varying strengths of pre-positioning, air mobility, seaborne logistics, and more traditional amphibious operations. Equipment and supplies have been pre-positioned in Norway for a specialized cold weather marine expeditionary brigade to be air-lifted in within days of alert, with thirty days of 'initial sustainability'. Supplies must, however, be *sea*-lifted from Trondelag up to the likely northern theatre of operations.[26] Secondly, there is a heavy mechanized marine expeditionary brigade that can be rapidly inserted and supported by the four ships of the Atlantic Maritime Pre-positioning Force. As the Joint Chiefs of Staff define their role: 'The MPS program is designed to combine the responsiveness of air-lifted marines with sea-lift of pre-positioned equipment'.[27] The troops are air-lifted 'to the vicinity of the objective area where they join up with the ships and their equipment.... This brigade ... can be combat capable and ready to move to designated objectives in as little as five days of arrival in the assembly area.'[28] Finally, the Striking Fleet brings in the rest of Second Marine Expeditionary Force in its Amphibious Striking Force Component, as well as the carrier air power and other naval support that would probably be crucial to eventual success.

In addition to American forces the UK/Netherlands Amphibious Force, a brigade of 6,500 men, also has reinforcing Norway as its primary role. It is rapidly available in theatre at a nominal week's notice. The force can be 'poised' as required, i.e.

> held in readiness ... rather than committed prematurely....
> Depending on the situation ashore and recent intelligence the decision on where to land can be delayed till the last moment (provided that contingency plans have been made for alternatives) ... the entire landing force while embarked can be moved a distance of up to 300 nautical miles in 24 hours, and does not require large sophisticated beaches over which to land.[29]

European naval forces, operating under land-based air cover would be required to support and escort this operation, either in addition to those committed to the Striking Fleet or as a prelude to moving westwards to join up with the main force.

Before the main Striking Fleet arrives allied maritime forces, probably very largely European, must engage in 'precursor operations', clearing the fjords of enemy mines and submarines. The Norwegian Navy will, of course, have a major role in coast defence, mine countermeasures, coastal convoy escort (as stated above, sea communications will be vital in moving even pre-positioned equipment), and generally providing the 'socket' into which outside reinforcements can be 'plugged'. Nevertheless, the Norwegians keep their options open. As Vice-Admiral Rein, the Naval Commander Northern Norway diplomatically put it: 'The uncertainties related to the time aspect of deploying NATO forces into the Northern area imply that our national forces may get heavily involved in crisis management, or even combat operations, before allied forces could arrive and provide the desired support.'[30]

The defence of Norway would be a joint operation between Supreme Allied Commander Atlantic (SACLANT) and Supreme Allied Commander Europe (SACEUR). The Shallow Seas Campaign brings in CINCHAN (Command in Chief Channel) also, and covers the Baltic, North Sea, Channel, and South West Approaches. In the Baltic the emphasis would be on local fast-attack craft, coastal submarines and aircraft, notably the Bundesmarine's powerful missile-equipped Tornado force. These light, relatively unspectacular forces of a three-dimensional kind have the vital task of safeguarding the Central Front's northern flank. In the North Sea and close South Western Approaches the emphasis is once more on MCM and ASW, and the safeguarding of both ships in the final stages of their oceanic voyages to Europe and those engaged in short sea passages transporting reinforcements from the United Kingdom to the Continent and (in the early stages) families and other civilians the opposite way. Again shore-based air, surface and sub-surface assets have to act in close co-operation, e.g. the use of the aviation training Royal Fleet Auxiliary *Engadine* as a forward base for shore-based ASW helicopters.

In the Mediterranean, SACEUR via the Admiral who is CINC-SOUTH has two maritime campaigns of his own, Eastern Mediterranean and Mediterranean Lifelines. Here the 6th Fleet has traditionally provided one of the NATO's core air-striking forces. This might still be important, although the carriers might well have other calls on their attention to the east as well as to the north. Amphibious operations might also play a part on the southern flank, either in the classic scenario in the defence of Turkish Thrace or elsewhere. It would also be necessary to use sea denial submarine, air and fast-attack forces, as well as minefields, to deny the Soviets any amphibious options, e.g. the Black Sea. Local navies would come into their own here.[31]

As for the Central Front itself, the NATO Striking Fleet has traditionally been allocated on air support role. This reflected NATO's early weaknesses in land-based air power, as well as the Striking Fleet's lack of targets in its early days once the Soviet cruiser or two that had broken out into the Atlantic had been dealt with. Nowadays the Central Front role seems less important, although it is still exercised. One suspects that now the US Navy has got a worthy opponent in the Soviet Fleet it will tend to concentrate its major surface assets on the latter's defeat, or at least in defence of major NATO assets from an enhanced Soviet threat. SACEUR will have to rely on much more appropriate land-based aircraft, leaving the carriers to cover targets to which they give special access. Nevertheless, if NATO air power does suffer catastrophic losses US carriers might well prove useful to plug gaps or to mount desperate nuclear escalations. In such circumstances sea-based cruise missiles might play a useful role.

So far the discussion has concentrated on a European scenario but it is quite possible, if not likely, that any coalition war would spread round the periphery of the USSR. The Maritime Strategy indeed welcomes the probability of horizontal escalation, the application of maritime pressure against the Soviet Union's all round periphery. The Far East PLARB bastion and Soviet Far Eastern naval bases are as

Navies in war

fair game as those in the Arctic, especially as one of the main roots of the current Maritime Strategy was the Sea Strike plan for 'a multi-carrier battle group operation directed against Soviet Pacific naval ports such as Petropavlovsk', prepared in the mid-1970s by Admiral Thomas B. Hayward (later CNO 1978–82).[32]

The maritime battle in Pacific waters would be of even greater long-term significance than that in the Atlantic, for it is here that the key industrial base of the capitalist world is increasingly concentrated. In a future protracted war between US and Soviet-dominated coalitions many of the main munitions industries of the US coalition would be in Japan, the leading OECD producer of iron and steel and motor vehicles, and to a lesser extent in South Korea and even Taiwan. This would require a great deal of shipping, both to take in fuel and raw materials, and to bring out finished products. Japan depends on external supplies for almost all of her iron ore and crude oil: she even exports 80 per cent of her coal. Some stockpiling has taken place in Japan but her near-total import dependence, and the need to continue to use the massive industrial facilities both there and in Asia, adds up to a formidable naval problem. One wonders how far the Japanese Navy, for example, has learned the lessons of its poor record in shipping defence operations in 1941–5. Barrier patrols can now be expected, given the very unfavourable Soviet geographical position, and there are few reasons for the United States to forego its predilection for a carrier offensive against Vladivostok or, more especially, isolated Petropavlovsk. Increasingly sophisticated Japanese vessels should be able to give assistance in coping with the air and ASW threats. Again, however, force of losses might well dictate something of a reallocation of assets to direct shipping defence duties as time went on. Once again, forward operations into the heart of the Soviet bastions in the Sea of Okhotsk are probably best left to submarines, leaving the US carriers to knock out the various isolated forward Soviet bases, e.g. Kam Ranh Bay or any other future North Korean facilities. Amphibious landings on Sakhalin, or similar islands, might well have a part to play in an American Far East maritime offensive. It must also not be forgotten in this context that even such an important Soviet base as Petropavlovsk on the Kamchatka Peninsula is effectively an island.[33]

A third front in a superpower war would be the Middle East, especially the Gulf area. Indeed, a Soviet American confrontation in this highly sensitive and economically important area might be the spark that set off the conflagration in the first place. The level of military activity in this region is crucially dependent on the level of US forces diverted from other areas, but the sea inevitably plays a major role in any US deployments. The US Central Command obtains a quarter of its allocated strength in fighter strike aircraft from US Navy carrier forces, both those in the Mediterranean as well as those in the Indian Ocean. Carrier-based aircraft would be absolutely vital to a US air effort if the Americans were finding it difficult in obtaining access to local bases. Carrier based A-6 (or A-12) aircraft could reach into Northern Iran from both the Eastern Mediterranean and the Northern Gulf, although the capacity of US carriers to operate in the latter waters at least is a little dubious given lack of sea room and potential mining threats. Such forces (plus USAF B-52s from Diego Garcia) would demonstrate US opposition to any Soviet moves on land, and be effective against Soviet airborne operations. For a complete air offensive against a full-scale land attack, however, land-based aircraft flying from friendly states in the area would be imperative.[34]

A divisional-sized Marine Expeditionary Force (MEF) is earmarked for the Gulf area, but most of its troops would probably come by air. Supplies for the MEF are carried in the three Maritime Pre-positioning Ship (MPS) squadrons. Only one MPS squadron is at Diego Garcia, the other two are in the Atlantic and Pacific, but it is considered that all three could be in the area within ten days, each carrying thirty days' supplies for one of the MEF's brigades. This, of course, assumes there are not other calls on MPS assets. Other PREPO (pre-positioning) ships are at Diego Garcia; they carry about 55,000 tons of supplies for the air force units initially deployed in the area.[35]

The aim, as in Europe, is to use sea, air, and land assets in optimum synergy. Until the deployment of the three MPS squadrons, the whole MEF was to have come by sea in amphibious lift, which would have taken between 3–4 weeks. Now both the MEF and the 82nd Airborne Division are fully deployable within fourteen days of mobilization. This, however, also requires extra airlift capabilities. Until all 50 C-5Bs were purchased by 1988, the 82nd took as long to deploy by air as the MEF coming by sea. Moreover, in initial deployment time, sea and air are surprisingly equal. It takes 48 hours both to deploy the initial battalion of the 82nd to an airfield in the Gulf area and to deploy a marine battalion from the Indian Ocean. The 7th Marine Expeditionary Brigade can be in the Gulf with amphibious ships in a week, by which time the 82nd would have been built up to a roughly similar level. Only then does enhanced airlift begin to give the advantage, and only on the basis of heavy equipment coming by sea.[36]

The purchase of the eight Algol class, 30 knot vehicle cargo ships allows one of CENTCOM's (Central Command) Army divisions to be in the Gulf with supplies within four weeks, about the same time it takes the USA's enhanced 1988 airlift capability to carry an infantry division there. Slower sealift brings in another Army division in about 35 days, as well as supplies to keep operations going.[37]

Whether such forces would be capable of successfully defeating a Soviet army with all the advantages of land communications and internal lines is another matter, but the terrain of Northern Iran is not very favourable to a rapid offensive.[38] After all, the major aim initially would be to demonstrate American interest and commitment. The sheer possibility of the deployment of forces of this size also must give the Soviets some pause for thought. They must calculate that the Americans might well oppose a thrust to the Middle East in a way that either could prevent any early conventional success, or that would provide the conditions for a nuclear war. Four or five divisions is a force that no US president could see destroyed without seriously considering nuclear escalation. The synergy of nuclear and conventional forces, like that of land and sea power, creates conditions where Soviet military adventurism is probably successfully deterred. The problem of forces and lift assets being diverted elsewhere by horizontal escalation is, however, significant. Marines fighting in Iranian passes cannot storm ashore on Sakhalin; fast cargo ships racing for the Gulf cannot simultaneously reinforce Europe.

Having some potential for action may be the most significant factor. For the dangers of nuclear escalation would be a constant concern in any conventional conflict on the scale just described, especially if one side or the other scored successes approaching the decisive. These dangers will probably continue to enforce the considerable caution in avoiding direct clashes that has generally characterized superpower relations to date. This will, however, still leave much scope for hostilities between smaller states, where the stakes are not so high, except for the immediate participants.

There have, of course, been many such 'limited' wars in the period since the end of the last great coalition war in 1945, sometimes with an external power intervening on one side to help. These conflicts emphasize what are often called 'power projection' capabilities of navies. Naval air forces and, to a lesser extent, naval artillery, have proved to be useful and flexible means by which external powers can bring fire power to bear against less sophisticated opponents. Indeed, from about 1950 to about 1973 and the end of the Vietnam War, this was the major war-fighting role of the West's navies.[39] Such firepower helped stave off Western defeat on a number of occasions, although it could not ensure victory. Smaller powers can also use sea-based fire power where appropriate, e.g. Israel against its neighbours. The limited nature of the opposition also makes amphibious landings in limited war even more possible than in higher-level hostilities, with potentially decisive results, e.g. against the North Koreans in 1950 and the Argentines in 1982.

'Sea power' can only be decisively and definitively influential, however, in certain circumstances. Unsurprisingly, when the aim is to capture or recapture an island, it can be the most important factor, but even in these cases other circumstances have to be favourable. In 1982, a few hundred miles or so of extra proximity to the Argentine coast might have made a great deal of difference. When the target is physically connected by adequate communications to a major land mass the problems are even greater. The Inchon landings in 1950 were suddenly reversed by a massive application of continental power in the shape of the Chinese People's 'Volunteers'. In major wars between Third World states (which given their ferocity are better termed 'local wars' than 'limited'), e.g. the Indo-Pakistan War of 1971 or the Iran–Iraq War of the 1980s, high levels of sea fighting can take place, but often only to limited effect. In 1971, Indian control of the waters off what was soon to be Bangladesh isolated the area from the rest of Pakistan, and made external military intervention troublesome. But the Indian victory might have conceivably taken place without any maritime operations at all. Only if the Pakistanis had been able to blunt the remarkable Indian blitzkrieg into the disaffected province and create a defensible bastion would sea power have been able to come into play, and this begs certain questions about Pakistan's capacity to mount a maritime reinforcement or relief operation.

In the war between Iraq and Iran there has been much more scope for the traditional application of maritime power. Iraq rapidly made herself much less vulnerable to this kind of pressure, being supported by her Arab friends and being able to export oil by pipe through Turkey and Saudi Arabia to help sustain her war effort.[40] Iran, however, is dependent on ships for her vital oil exports, and these were vulnerable to Iraq's 'naval' forces in the shape of missile-firing aircraft.

Although ships had been attacked by Iraq previously the campaign began 'in earnest'[41] in Spring 1984. The Iraqis were hobbled by limited capabilities in aircraft range, reconnaissance support, and lack of missile effectiveness against mercantile targets, especially large tankers. The Iranians used convoys to protect their shipping and introduced a shuttle tanker system requiring large tankers (some VLCCs), equipping them heavily with ECM equipment and running them down to new trans-shipment terminals around Sirri and Larak at the southern end of the Gulf. Only slowly were the Iraqis able to acquire the aircraft range to attack this lifeline effectively along all its length, and it took two years before the Iranians began seriously and significantly to be hit by the attacks.[42]

The Iranians began retaliation against this Iraqi *guerre de course* in 1984 using aircraft-carrying weapons like the Maverick electro-optically homing missiles, and even Sidewinder air-to-air systems.[43] As well as being even less effective than those of the Iraqis, Iranian attacks were also much less numerous until the pressure of 1986 caused significant escalation by Iran. In 1985 there were forty Iraqi attacks on shipping and only thirteen Iranian: in 1986 the figures were sixty-five and forty-one respectively.[44] Iran also began to make much greater use of surface forces, including her Sea Killer equipped frigates. She also began to acquire shore-based anti-ship missiles from China and perhaps most significantly began to deploy mines from small motor boats.[45]

Iranian strategy seems to have been to exert pressure on Iraq's Arab allies, notably Kuwait. It also, however, had the effect of widening the naval involvement both of the superpowers and of the Western Europeans. All these powers had significant interests in maintaining freedom of shipping to the Arab States, and the stage was set for the increasing naval deployments and reflaggings of 1987.[46] The involvement of outside naval actors, however, illustrates another significant factor in limiting the role of the sea power of the participants in local wars. Outside powers with an interest in maintaining the flow of maritime trade, especially in vital materials such as oil, may well not let local sea power prove more effective. Three factors then, (a) limited vulnerability to maritime pressure, (b) limited capability to inflict it, and (c) the likelihood of intervention by outsiders, given the international nature of the maritime environment, all militate against major *local* naval impact in the majority of wars between Third World states.

It is conceivable, however, that local navies might play more of a part as Third World states develop their maritime interest and their forces. One or two maritime conflicts might escalate into situations where force was used and replied to on a high enough level, and consistently enough, to be characterized as war. The main area of potential naval conflict probably lies in South East Asia, where conflicting areas abound in an area ripe for hydrocarbon exploitation. A more favourable economic context for these activities could possibly, with other factors, increase the stakes to the point where one of the non-possessing claimants felt impelled to use force, as when China seized the Paracels in early 1974: a counter-attack might lead to more sustained hostilities. But even here the probability of the USA, USSR, and China becoming involved would probably act to limit the extent of hostilities.

Once outside actors do appear, their impact on a highly sea-dependent participant in a local war can be very important. The escalating activities of the US Navy in the Gulf in 1987 and 1988 clearly showed Iran that there was nothing she could do to protect her vital oil exports should the USA decide to interdict them. Mining failed to drive US forces away and when the Iranian Navy and Revolutionary Guards finally gave battle following the retaliatory attacks on the oil platforms at Sirri and Sassan on 18 April 1988, the result was the loss of a frigate, a fast-attack craft and two small gunboats, and damage to a second frigate. A shore-based Silkworm missile was easily decoyed away.[47] Iranian naval impotence had been clearly demonstrated. In combination with other factors – the missile 'war of the cities' and reverses on the battlefield – outside maritime pressure (added to Iraq's demonstrated capacity to devastate Iran's far-away oil terminals) obtained Iranian acceptance of Resolution 598. As ever, sea power did not work alone, but it must have been a key factor, given the high stakes. The US Navy was threatening Iran's very economic survival.

Thus, although they may not actually participate in the fighting, navies can play a highly significant, if not crucial, role in limited or local wars. Their presence can act to set the boundaries for a conflict by demonstrating the willingness of larger powers to intervene if things get out of control. They may be used to exert pressure on one or both sides to accept a ceasefire, especially in a protracted conflict in which neither side can clearly win. Similarly, the presence of one major power's naval forces may neutralize the activities of the other by the demonstration of the potential capacity to disrupt any planned intervention. The paradox of this situation is that the very presence of the outsider's naval forces may well be the major factor in preventing the local navy exerting greater influence on the outcome of the war itself. Military sea power, therefore, will continue to play a key role in the maintenance of peace and security, just as the civil uses of the sea will continue to be crucial to the world's economic prosperity and well-being. The connection between the two will, however, be ever more subtle and complex. Can one produce some new theory to provide a framework for thought in this developing situation? It is to this task we shall finally turn.

Part VI
Conclusion

Chapter 11

The future of sea power – three theoretical frameworks

This final chapter will attempt to develop three theoretical frameworks that it is hoped will assist those who wish to think about sea power over the next few decades. None are very original: indeed all three are based on existing ideas, albeit amended in significant ways. The first will cover the foundations of sea power, why nations will seek to maintain naval forces and other capabilities to exert power at and from the sea. The second will cover the roles of navies, what they will seek to achieve at sea, how they will do it and how those tasks will change in emphasis in the future. The third will try to construct a typology for navies, putting them into different analytical categories.

The foundations of sea power

Mahan is a name often taken in vain by commentators on sea power; in vain, because they sometimes reveal a certain lack of knowledge of what he actually wrote. Some contemporaries, notably Corbett, doubted the American's depth of analysis at the time and, as we saw at the outset, his fundamental thesis has been seriously questioned since.[1] Nevertheless, given its seminal nature, Mahan's 'Discussion of the Elements of Sea Power', Chapter 1 of *The Influence of Sea Power Upon History*, which provides the basis of his reputation as the leading analyst of the subject, deserves close examination in the context of developments since he wrote it and likely development in the future.[2]

The sea remains, as Mahan wrote, 'a great highway . . . a wide common over which men may pass in all directions'.[3] Despite the growth of other forms of economic sea use (see Chapter 3) and the concomitant creeping assertion of greater sovereignty and/or jurisdiction over what were previously high seas (see Chapter 7), it seems unlikely that the principle of the free passage of merchant shipping (if not warships) will be significantly constrained. Indeed, the wider

Conclusion

distribution of merchant shipping and the likely increase in world sea-borne trade will give more and more states an interest in allowing it freedom of transit. As in Mahan's day, it will still be easier and cheaper to transport heavy and bulky items by water.

Where Mahan is on less firm ground, however, is his bland assertion that 'it is the wish of every nation that the shipping business should be done by its own vessels'.[4] This is clearly *not* the case in the contemporary era of economic liberalism. As we have seen, some countries, notably the United Kingdom, show surprising carelessness in the fate of their merchant fleets, choosing to pay the price in balance of payments losses rather than the cost of maintaining ships at higher than 'market' rates through subsidy or other support. It is too early to say yet whether states such as Britain will appreciate the full implications of this policy before terminal damage is done to their national capacity to transport goods by sea in crisis or war. The Government argues that there will be enough available ships for such circumstances. Others – especially interested parties – are far less sure and argue for special measures to maintain a core fleet.[5] One suspects that most states will take special measures to protect *some* minimum seaborne carrying capacity for a combination of strategic and economic reasons. Even that bastion of free market economics, the United States, has little hesitation in doing so and has even made 'sealift' a naval priority. It will all depend, as Mahan himself pointed out, on 'the character and clear sightedness of rulers'.[6]

One crucial factor affecting governmental attitudes towards merchant fleets will be Mahan's 'spirit of the age',[7] the general trend towards or away from protectionism. The signs are currently ambiguous. Greater protectionism will, however, imply more measures, in certain countries at least, to protect mercantile marines from foreign competition. A major break-down of the current liberal world economic order would probably also reforge some of the broken links between military navies and mercantile marines. If states are locked in overt and bitter economic competition, merchant ships may well depend, as they did in the age of classical mercantilism, on 'the protection of their country throughout the voyage'.[8] Yet, despite danger signs, the trend, in both the western world and the Soviet bloc at least, seems to point more firmly in the direction of greater interdependence and economic liberalism. In this context, as now, reliance for naval protection will have to be placed on certain states' more sophisticated analysis of the need to protect freedom of navigation as an act of enlightened self-interest for themselves, their allies, and their trading partners, and for the whole economic system from which they benefit. The states which carry out this duty may well not have large mercantile marines but their interest in overseas trade and general good order will be great nevertheless.

That naval power might be powerful globally, regionally, or just locally. The potential mismatch between flag of merchant ship and flag of warship may, for the time being, create certain legal difficulties. It may, perhaps more importantly, prevent the interest in defending a particular trade being perceived. But the evidence of recent events in the Gulf demonstrate that warships will, eventually, come to the aid of all merchantmen threatened in a particular area, e.g. by combatants whose belligerent rights are in dispute. States who are the beneficiaries of maritime trade – albeit carried in other peoples' ships – will deny themselves the capability to do something in defence of that trade at their peril.

In the future multipolar world, it seems likely that the responsibility to protect shipping in peace and war will be spread more widely. Local powers will increase in importance, powerful regional actors like India will emerge, and the growing great powers of Japan and China will probably match and surpass the existing medium naval powers of Britain and France. Despite all the problems, the latter will almost certainly *eventually* combine together with their Community partners into some kind of loosely constructed European naval force. It seems unlikely, however, that any of these states will require what seemed a vital attribute of distant water operations in Mahan's day: colonies and foreign bases. Not only are formal colonies now unnecessary to safeguard economic contacts, but the striking feature of sea power in the latter half of the twentieth century is its independence of permanent shore bases. The United States and Soviet Union both have some 'bases' but these are forward outposts for fleets largely maintained from home dockyards by means of afloat support. Admiral Hill's vital attribute of 'reach' is inherent in any navy willing to invest in afloat support, tankers, and repair ships.[9] Local friends will, in all likelihood, provide harbours in which such support vessels can reside (as the Gulf states quietly gave the European navies house room in 1987–8).

Mahan summed up the keys to sea power as 'production, with the necessity of exchanging products, shipping, whereby the exchange is carried out, and colonies which facilitate and enlarge the operations of shipping and tend to protect it by multiplying points of safety'.[10] All these 'keys' are now internationalized to a degree that Mahan would have found surprising and even shocking. Asia now produces cars, steel, and consumer electronics for the world. Huge merchant fleets now carry the flags of minor African and Asian states. Almost all the ex-colonies now have at least national sovereignty and seats

Conclusion

at the United Nations. This maritime background still creates interests that states perceive as requiring defence by military force deployed on, over, and under the oceans. But that perception requires and will require even more of the aforementioned 'character and clear sightedness'[11] from governments than that required in Mahan's day.

Mahan argued, deterministically, that the 'shrewdness and foresight of governments' had been less important in determining 'the history of seaboard nations' than certain 'natural conditions' that shaped their development. These 'principal conditions affecting the sea power of nations' were: 'I Geographical Position. II Physical Conformation, including as connected therewith, natural productions and climate. III Extent of Territory. IV Number of Population. V Character of the People. VI Character of the Government, including therein the national institutions.'[12] Let us critically examine these in turn from the perspective of the late twentieth century and beyond.

Geographical Position

Mahan argued that 'a nation . . . so situated that it is neither forced to defend itself by land nor induced to seek extension of its territory by way of the land . . . has, by the very unity of its aim directed upon the sea, an advantage as compared to a people one of whose boundaries is continental'.[13]

The invention of the aircraft and the rocket, which Mahan could perhaps not have foreseen back in 1890, have enforced strategic unities between continental land masses and offshore islands that have done much to undermine this basic Mahanian tenet. Britain, for example, can ignore the occupants of the coast twenty miles across the Channel less than ever. As one of the wisest First Sea Lords put it in relation to the commitment of land and air forces to the continent of Europe in March 1950, a Britain facing a hostile continent in modern conditions would be 'menaced as never before in our history'.[14] Nations, however insular, require defences and/or deterrents against air attack (which might, of course, be seaborne), and often armies to keep the enemy as far away as possible, at least as much as they require navies for traditional tasks. Geography is now much less of a defence from the need to weigh strategic priorities. The world is in many ways a much smaller one. Few, if any, states, therefore, can now emphasize their navies at the expense of other forces because of mere geographical position.

Mahan regarded the position of his country on the two oceans as 'either a source of great weakness or a cause of enormous expense'.[15] In fact it is an enormous American advantage given an ability to mobilize huge resources. The United States can deploy its navy easily into either ocean, Atlantic or Pacific, and has no need to defend itself on land from its weak neighbours. The other superpower, on the other hand, must deploy its fleets in widely separated areas behind choke points easily defended by hostile powers and must pay due regard to the overriding need to protect its land frontiers and maintain order in its western buffer. Mahan need not have worried.

Physical Conformation

Mahan argued that countries with long seaboards and good harbours tend to be more maritime than others. Richard Hill's evidence on sea dependence in Chapter 8 would seem to bear out this point. It is, of course, not just the possession of a coast that is important. The Soviet Union has one of the largest coastlines in the world but most of it is iced up and effectively unusable in terms of access except at great technological effort. The fact that the Soviet Union's harbours are relatively few, and sometimes isolated, is also a source of potential vulnerability to maritime pressure.

Islands are obviously especially sea dependent: 'When the sea not only borders, or surrounds, but also separates a country into two or more parts the control of it becomes not only desirable but vitally necessary.'[16] The level and extent of that control is, however, a function of other factors.

Extent of Territory

A fundamental weakness of Mahan was that he was unconcerned with 'total number of square miles which a country contains'[17] in his analysis of a nation's potential as a sea power. In this he was almost completely wrong. True, possession of large tracts of unexploited territory on its own is no help, but the current century was to prove that the extent of territory in terms of masses of exploitable resources was a key factor in building state power to equally massive proportions. Mahan clearly felt ill at ease with the continental preoccupations of his own time and country: he did not realize that his compatriots were laying the foundations for the largest expansion of military sea power the world has ever seen. In the modern and future world of more easily exploitable land masses, large territory often means more economic power if the wherewithal is there to unlock it. More economic power in turn means more sea power.

Conclusion

Number of Population

By this Mahan meant the number of the population 'following the sea, or at least readily available for employment on ship board',[18] and related it to his point about the need for coasts and harbours. Again he was wrong and for much the same reasons. The nature of modern navies is such that the overlap with other forms of maritime activity is relatively small. Military naval officers see themselves as separate, perhaps increasingly separate, from other sea users. Thus a large population 'following the sea' is not necessarily a prerequisite for an ability to deploy military power on the oceans. Even in Mahan's own USA, most naval officers come from deep inside the American continent.

This is not to say that having little in the way of seafaring population does not have implications in terms of the reserves that can be called upon to sustain military navies in crisis, the perceived importance of general maritime activities (of which military navies are a part) and the general status of a country in maritime affairs. But the situation is not as Mahan baldly stated it. Large numbers of suitable people may be important if there is a naval requirement for them, just as large quantities of any other resource might be significant, but they must be suitable for naval employment. This has much more to do with the general skills and education of the population than it does with any physical or professional association with the sea. Fishermen, or even merchant seamen, may need as much training as landsmen in order to operate a modern warship. The overlap between naval skills and other maritime skills is nowadays surprisingly limited.

Finally, and most importantly of course, the number of population in its relationship to the general resources and economic development of the state is a vital factor in determining the resources available for diversion to naval purposes. Countries with very large populations which they can only just feed are going to find it difficult to deploy powerful navies. This is, however, not just a question of diversion of resources into naval channels. It also reflects the resources available for education and training, and general industrial development. This leads us on to Mahan's next point.

National Character

Here Mahan was at his most 'nineteenth century' in outlook. He detected a natural 'tendency to trade, involving the production of something to trade with' as 'the national characteristic most important to the development of sea power'.[19] Our appreciation of

Three theoretical frameworks

economics now goes somewhat further than instinct, although cultural factors do play a part in economic growth and power. There seems to be good reason, looking at the comparative performance of the USA and USSR, to see certain factors beyond the material in the degree of economic success enjoyed by the two states. Britain and Japan also make interesting comparisons. However, the analyst is on firmer ground if he sticks to more easily measured and explained phenomena.

It is probably better to divide Mahan's old 'national character', like Gaul, into three parts: Economic Strength and Technological Prowess, as well as the more diffuse Socio-political Culture. The first is measured in GNP, the second in the quality of the technological output of the state, the third is harder to quantify but clearly visible in certain notable cases.

Only countries of considerable economic strength can afford to be major sea powers in the modern world as the cost of naval equipment of all kinds increases. As Chapter 8 has shown, sea power is a direct function of economic power. Even a country like Japan, with considerable inhibitions about mobilizing its enormous economic strength, cannot help but deploy one of the world's most significant navies. As touched on above, it is not just a question of absolute strengths. Countries with low per capita GNPs, like India and China, are currently relatively weak sea powers for their size. Their sea power will grow as their economies get richer and they bring their populations under control. Countries with high per capita GNPs, on the other hand, may already be sea powers of some importance. They have the resources to invest in warships.

Technological prowess, however, is of equal significance. A well educated, technologically literate population is essential both to man and fight modern warships, and produce the equipment that allows them to fight. The modern warship, or other naval platform, like the ship of the line of the nineteenth century, is often one of the most sophisticated items of contemporary equipment made by a nation's industries. Its weaknesses often reflect the general technological weaknesses of its owners.

The third factor, socio-political culture, is historically rather than ethnically based. Certain states have, for various reasons, eschewed traditional naval expansion. The two classic cases are Japan and the Federal German Republic. Both have been considerable sea powers in their time and both could be greater sea powers than they are today if they chose. Japan, especially, is perhaps the world's greatest maritime nation and as world number two economically should be a major naval power. That she is not is a product of psychological and political inhibitions firmly implanted by the traumas of defeat. The

Conclusion

Federal Republic is in the same state. The movement of its warships to the Mediterranean in late 1987 to sustain NATO's strength there was hailed as a very special event. It was made clear that these ships could go no further 'out of area'. West Germany is still not that sort of country.

These inhibitions are fascinating, not least because they are self-imposed. The United States often thinks of trying to make its allies do more militarily, although those with a feeling for the past might be grateful for a quiescent Japan and Germany. As self-imposed limitations they could well change over time, depending on the external environment. Japan's self-denial has been part of a wider unwritten trade-off that allowed her, unlike before 1941, virtually unlimited penetration of markets and raw material sources in a liberal world economic order. This condition need not last. If the protectionist factors mentioned above prevail and Japan was suddenly faced with a set of closed trading doors the effects might be both surprising and terrifying for those with short memories.

The Federal Republic is not the only country in Europe to feel inhibitions about acting navally on a wider stage. The Netherlands has reacted to centuries of imperial decline by a considerable ambivalence towards foreign entanglements. Even when her energy supplies were threatened in 1987, the context for sending naval forces to safeguard them had to be international if political consensus was to be maintained. This is probably a key to the future: a greater *European* rather than *national* dimension to naval activity might re-legitimize and modernize uses of naval power at some distance from Europe's shores when European interests are threatened. By such means the economic power and technical prowess of Europe as a whole might be applied more powerfully and extensively at sea.

A final aspect of this factor relates more closely to Mahan's original concept, the effect a nation's socio-political structure has on the qualities of initiative, resourcefulness, and independent thought required for naval operations. Although this can be overstated, as all naval forces must operate in a disciplined, organized and controlled manner, the stereotyped operations of the Soviet Navy compared to Western navies has some relationship to the negative and restrictive aspects of Marxist Leninist society. This brings us to the last of Mahan's factors.

Character of Government

Mahan doubted 'whether a democratic government will have the foresight and the keen sensitiveness to national position and credit, the willingness to insure its prosperity by adequate outpouring in time of peace, all of which are necessary for military preparation ...'[20] However, as Philip Pugh has shown, the rise of democracy in Britain has not seen any decrease in the defence burden in peacetime. It has fluctuated between minima of 2 per cent and maxima of about 5.5 per cent of GNP in every period of peace from 1698 onwards. Indeed the figures for 1969–83 are the highest minimum over the three-century period.[21]

Government policy can foster sea power by fostering general economic growth, or by specially emphasizing defence in general and the navy in particular within the normal peacetime margins. This can be a feature of governments as different in ideology as the Reagan Administration or Brezhnev's Politburo. On the other hand, equally right-wing governments in the West to President Reagan's may allow a devotion to free market competition to destroy the 'industries and sea going bent'[22] that gives a country those interests at sea to defend that Mahan thought to be the only legitimate basis for a nation's sea power. Britain under the Thatcher Administration is a classic example. As military sea power operates nowadays at least semi-autonomously from 'peaceful commerce', however, the latter effect may not be decisive in reducing a nation's capacity to exert most aspects of military power at sea.

Government policies can affect to a considerable extent a nation's capacity to mobilize extra naval assets from the merchant marine, and sustain military operations at sea by the same means. Nevertheless, it is an open question whether any resultant reduced perception of 'maritimeness' affects first-line military naval capabilities very much. This author suspects it might not, at least in the short to medium term. Other factors, for example wider national perceptions of security interests, usually have a greater impact, for example the need to contribute to the overall military posture of an Alliance with a superpower patron. Shipping and shipbuilding are not the only industries with a security relationship deserving of government support when required and as appropriate. Their defence relevance is, however, more direct than most.

What are we to make then of Mahan's determinants? They need both re-ordering and changing. At the top of any list of elements of sea power must go Economic Strength. This is the foundation upon which all stands. It comprehends the old 'extent of territory' and 'number of population'. Large countries with numerous populations and powerful economies can devote resources to military sea power, as sheer expressions of that strength, *even if they have little other interest in using the sea*. The economic strength of smaller states also fundamentally affects the level of sea power they can afford, but it is not just a question of sheer resources. They must be of the right

Conclusion

kind, be put into the right form, and deployed with a certain lack of inhibition.

This leads on to Technical Prowess. A country with considerable economic strength may not have the capability to build a high technology navy and operate it effectively. The Soviet Union has had considerable difficulty in building up an effective fleet. Even after half a century of industrialization, the USSR still lags behind the USA at sea. Much of its technical development has been derivative, reliant on technology transfer and even espionage. Although the United States, and to some extent Western Europe, will retain a lead in naval technology it will slowly be eroded. More countries are beginning to build warships and even quite sophisticated naval weapons are being produced by countries that were previously reliant on the established naval powers for equipment. China now produces quite sophisticated sea-skimming anti-ship missiles and puts them on submarines of her own design. Taiwan also produces SSMs that are copies of Israeli weapons. Brazil is developing her own SSM. The trend towards wider production of naval technology will continue.

There is an interaction between technological prowess and the next 'condition', which is Socio-political Culture. Certain societies and political systems are better at adapting to technological change than others, but even when, like Japan, they have demonstrated a quite remarkable adaptability and prowess they may still significantly limit their investment in the naval dimensions of security and foreign policy. Others countries may also feel similar inhibitions. This seems to operate almost independently of the sea dependence of that state. Equally, countries that are not overly sea dependent can feel the need to use the seas to express and exert military strength, both offensively and defensively (although they will probably interpret the former in defensive terms).

The above three conditions seem to be of a first order. Even large islands with good harbours are of little naval significance without people, effective ships, or aircraft. But the second order, more specific, conditions – those affected by national situation – are none the less important. First comes geographical position. Land-locked states will tend not to have significant navies, although it must be admitted that water is such a universal and important transport medium even within continents that Switzerland, Hungary, and Bolivia all appear in the pages of *Jane's Fighting Ships*.[23] As we have seen above, sheer position is of less significance in today's smaller world. Other geographical factors are perhaps best comprehended in Admiral Hill's wider concept of Sea Dependence with its various criteria, seaborne trade, merchant marine, shipbuilding, fishing catch, and off-shore zone.[24] Countries with a high degree of sea dependence will have more reason to deploy coercive power at sea than others, although whether they actually do so will be affected by the first order conditions.

This leads on to a final second order factor. Character of Government should be changed to Government Policy and Perception. The way the other factors are mediated by governments reflects the specific constellation of political forces in a country at the time. Governments may ignore certain interests and emphasize others. Governments may be elected or come to power in more violent ways because of historically specific circumstances. Governments cannot change naval policy very rapidly, given the long-term nature of naval forces, but they can affect how it is used in the short term and take decisions that may affect future sea power assets in significant ways. Government policy and perceptions thus have an impact of their own on a nation's sea power that is at least no more dependent on the others than the rest of the members of this highly interdependent list of factors.

Our new set of Principal Conditions Affecting the Sea Power of Nations thus becomes:

First Order
1. Economic Strength
2. Technological Prowess
3. Socio-political culture

Second Order
1. Geographical position
2. Sea dependence in terms of:
   (a) Seaborne trade
   (b) Merchant marine
   (c) Shipbuilding
   (d) Fish catch
   (e) Offshore zone
3. Government policy and perception

Putting sea power into these terms deals with one of the major problems of Mahan's theory. Mahan insisted that 'purely military sea power' can only be built up by an aggressive despot. Otherwise, 'Like a growth which having no root' it 'soon withers away.'[25] The history of American sea power over the last century would cast considerable doubt on this assertion. A perception of legitimate interest in how the international politico-economic system should be ordered, indeed a sense of duty that it had to intervene as the senior member of a set of like-minded nations, has impelled the largest democracy in the world to send the world's most powerful navy to sea. At times it has

Conclusion

seemed that a mercantile marine has been fostered to 'legitimize' this military sea power in a strange process of turning Mahan on his head. Other demands, military security and the protection of economic, political, or even ideological interests across the globe, will impel other nations too to use the seas for military purposes. Attempts somehow to legitimize a military navy in a large national mercantile marine are usually doomed to failure and are simply unnecessary.

Yet Mahan might have been right, in one way, to draw some attention to the relationship between the economic uses of the oceans and the military power deployed upon them. As the economic forms of sea use become ever more internationalized so the military uses of the sea may have to become more internationalized too. Perhaps a mismatch is even now arising, one that Mahan in a sense foresaw. In 1987-8, *national* navies got themselves into something of a tangle trying to defend *international* shipping in the Gulf. It took time for the situation to sort itself out. Is it too much to hope that as the superpowers become ever more aware of their common interests, the security of each other, and that of the other nations of the world, a new attitude to sea power may eventually be born; one that emphasizes international duties to safeguard an international maritime community. If it does, Mahan may have been proved right after all, but in a way that the old imperialist and mercantilist would never have imagined.

The roles of navies

Mahan's analysis concentrated on the use of the sea, albeit in the limited sense of a transport medium, and this leads on to a wider theory of the functions of naval forces. The 'use of the sea' was the basis of Ken Booth's perceptive triangular or 'trinity' model published ten years ago in *Navies and Foreign Policy*.[26] Booth rightly saw the vital factor being 'the use of the sea'. A state might wish to use the sea for three purposes: '(1) for the passage of goods and people; (2) for the passage of military force for diplomatic purposes or for use against targets on land or at sea; and (3) for the exploitation of resources in or under the sea'.[27] Booth's ordering of these priorities reflected the traditional Mahanian emphasis on shipping. The nature of the current and future strategic and economic world, however, notably the continued sea-basing of the most secure nuclear deterrent systems, would advise placing 'the passage of military force for diplomatic purposes or for use against targets on land or at sea' at the head of the list, with economic shipping next and resources last, but increasing slowly in importance. Booth then went on to construct his 'trinity' of naval functions:

The unity (the one-ness) of the trinity is provided by *the use of the sea*: this is the underlying consideration in the whole business of navies. . . . The character of the trinity is then defined by the three characteristic modes of action by which navies carry out their purposes: namely the military, the diplomatic and the policing functions.[28]

Deciding which category to place activities is trickier than it first seems. For example Booth, as we have seen in Chapter 9, separates diplomatic from military at the 'actual employment of force'.[29] Yet on his own definition the *latent* as well as the actual application of force comes under the military category. Equally, the constabulary duties of navies must include some recourse to the actual *use* of armed force, albeit limited at all times. This forces one, like Cable, into a distinction between higher and lower level operations, peace and war. No such definition is truly satisfactory, especially in a world where war is officially illegal under the UN Charter, and every conflict since 1945 has been undeclared and justified as 'self-defence' under Article 51.

Nevertheless, such a distinction has to be made. Strategic deterrence is not quite the same thing as gunboat diplomacy or showing the flag. The distinction must surely be, as Cable argues, at the point where the infliction of damage, actually carried out *or just threatened to one's opponent*, becomes an end in itself, be it in a limited theatre or more generally.[30] When on 2 May 1982 a provoked Britain, with her forces on the spot severely menaced, altered its rules of engagement to allow general attacks on Argentine warships, that particular rubicon was crossed: gunboat diplomacy became war. Clear preparation for such hostilities or worse, including nuclear city-busting, and the perception of these preparations in the context of a potential 'war' if certain boundaries (physical and legal) are crossed, is a rather different matter from using the capabilities thus acquired for specific and limited political or law enforcement purposes, even if the result of the latter use is the actual killing of someone.

Now, on this basis, we can reconstruct the Booth triangle. At the base go the 'military' roles, for which most naval forces are primarily designed. This can be divided in relatively traditional terms: power projection, sea control, and sea denial. Naval missions such as strategic deterrence (power projection), coast defence (sea denial), *guerre de course* (sea denial), and defending shipping (sea control), can be subsumed under one or more of these headings. All these objectives require distinctly military forces; they are concepts of maritime *strategy* properly and narrowly defined.

Conclusion

The 'diplomatic' role of navies can also be divided, *à la* Cable, into 'gunboat diplomacy' and 'showing the flag' (for the distinction see Chapter 9), and operate in a context, by definition, of inter-state diplomacy, be it on a unilateral or multilateral basis. If the context is some *universally agreed order* in which 'enforcement' is considered to be legitimate then one moves into the constabulary role (preferred to Booth's 'policing'). Currently this role is almost entirely national. International constabulary roles await the development of forms of enforceable and mutually agreed international law and order similar to those claimed by sovereign states over their own territories, territorial seas, and (within closely defined but expanding boundaries) exclusive economic zones. Figure 11.1 shows the 'use of the sea' triangle.

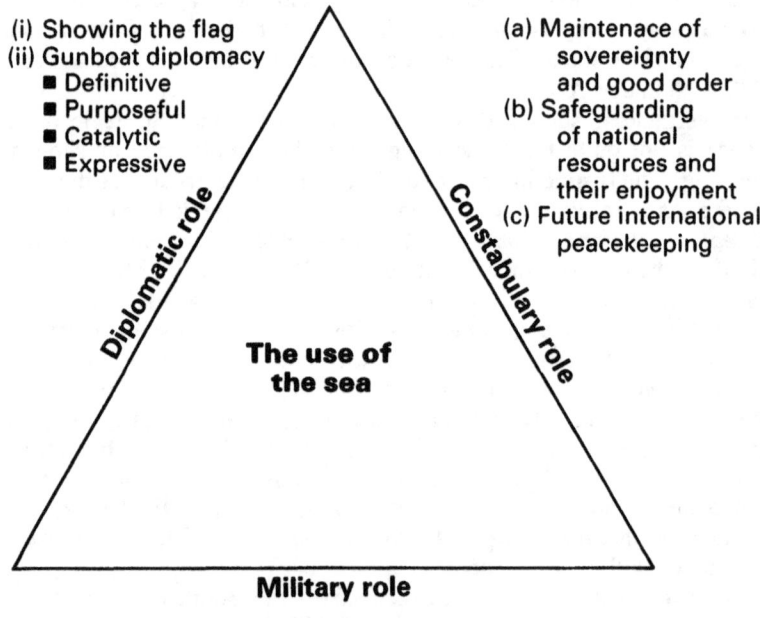

Figure 11.1 The 'use of the sea' triangle I

Three theoretical frameworks

Before leaving this, perhaps a few more words are necessary on 'sea control'. All that is meant by this term is the ability, when required, to deploy armed force and/or fight a maritime campaign to enable one to pass shipping defined in the widest possible sense for a specific purpose of one's own. By definition the expected opposition is serious enough to put that shipping seriously at risk. Sea control operations are often a prerequisite of some forms of power projection but certain types of forces may have what might be termed 'inherent sea control', e.g. silent submarines or very powerful warship groups facing limited opposition. Power can thus be projected over an 'uncommanded sea' without engaging in any greater sea control operations than submerging or providing a powerful enough escort that can brush weak opposition aside. Technology is, however, emphasizing the need for powerful close defences around assets at risk. As platforms get stealthier and weapons more formidable, the emphasis in military operations at sea is going to swing much more to direct protection of the form of sea use being contested. As one of the Ministry of Defence's key forward thinkers summed it up at the end of 1988, defence will have to be concentrated much more about the 'key sea users', whatever they may be.[31] Having moved away from Corbett's 'normal' we now seem to be moving back to it. (see Chapter 2)

Navies will continue to be used in these roles as they have in the past, although the balance of the contexts in which they work may change. One could extend Booth's triangle, which is already based on contexts, into an interlocking set of circles.[32] For NATO and Warsaw Pact navies, at least, these three circles would be: (1) East–West confrontation, (2) national interest, and (3) law and order. If the diameter of each circle is very arbitrarily taken to be (a) military role, (b) diplomatic role, and (c) constabulary role, we end up with a pattern like Figure 11.2.

Sometimes a particular naval event finds itself in one circle, sometimes at the overlap of two, sometimes all three. The precise location of any point is sometimes more a matter of interpretation than clear definition but the three circles diagram is interesting in its future implications. If East–West confrontation is indeed withering away, albeit very slowly and uncertainly, there may well be significant implications for naval forces. Certain military roles, e.g. forward power projection against superpower opposition, will be reduced in importance; others, like gunboat diplomacy and maintenance of good order, may well become more important. Forces most relevant for high-level operations may seem less important, especially nuclear-powered attack submarines, fleet carriers, and sophisticated ASW and AAW ships. Presence, perhaps at some

Conclusion

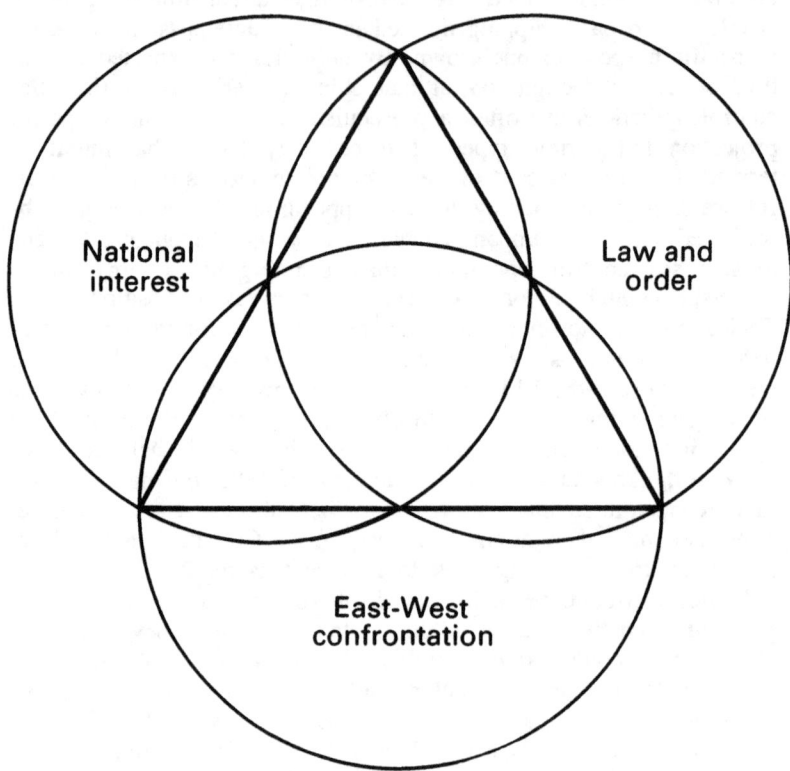

Figure 11.2 The 'use of the sea' triangle II

considerable distance, and the capability to deploy sufficient but limited force to make one's mark in a basically 'peaceful' context may well become the essence of naval power. In such activities surface 'warships' could be more important than ever. 'Sea power' in a very traditional sense thus may have a secure future in the longest of terms.

A typology for navies

A significant literature has grown up on how to rank navies in terms

Three theoretical frameworks

of their overall power and capability.[33] This is a difficult task as it must take into account factors such as the types of forces deployed, the sophistication of their equipment and level of afloat support as well as mere numbers of vessels. The categorizations below originate in Morris's work on Third World navies, as amended by Haines, but they try to go further to produce a 'global naval hierarchy'[34] that will form the basis of some speculative remarks on the future balance of naval power. Note that 'navy' includes all those forces capable of exerting force at sea, not necessarily just those bureaucratically organized into a 'navy'. Also *'force* projection' implies a capacity to engage in 'sea control' and 'sea denial' as well as *'power* projection'.

Rank 1: Major Global Force Projection Navy – Complete

This is a navy capable of carrying out all the military roles of naval forces on a global scale. Currently, only one nation's navy at this level of capability qualifies, the United States, and it will maintain this lead for some time to come. No other navy is likely to equal the capacity for force projection provided by the numerous carrier battle groups supplemented by a powerful SSN force. Surface action groups and convoy escort groups back up the carrier capabilities, while a sea-based ballistic missile force makes the main contribution to the USA's strategic air power. Amphibious and afloat support capacities remain unequalled and sealift capacity is being safeguarded and enhanced. The lessons of the Gulf have led to neglected MCM capabilities being improved. It is unlikely that any rival will seriously approach the United States in its ability to deploy sea power by moving into the first rank: the main danger to US hegemony is that she will decline in relative power to Rank 2 in which there may well be other members.

Rank 2: Major Global Force Projection Navy – Partial

Currently there is only one nation at Rank 2: the USSR. Although she deploys powerful SSBN forces and highly capable sea denial forces of oceanic range in the shape of submarines and shore-based strike aviation, the USSR does not have the ocean-going sea control and power projection assets to dominate the local naval environment far from her shores. Her cruiser-based forces will not provide her with anything approaching US capabilities in this regard. A limited number of specialist vessels and state-controlled merchant shipping provide supplementary amphibious and sealift capacity that can carry out limited power projection duties, but they do not approach

Conclusion

America's specialized amphibious fleet in scale or sophistication. Current trends seem to imply a reversion to a priority for adjacent force projection, but the USSR will not lose a limited capacity to be a significant naval actor on a global scale if she so chooses. As other nations approach the USSR's level of economic strength the latter may find herself with rivals at this level. The USA may also be reduced to it in the long term if she faces serious economic difficulties.

Rank 3: Medium Global Force Projection Navy

Currently there are only two candidates for the Rank 3 category: Britain and France. Both possess, to varying degrees, carrier and amphibious capabilities, sea control surface forces, SSNs, SSBNs, and afloat support. Each would be capable of conducting one major 'out of area' operation and each (Britain currently much more than France) would be capable of engaging in high-level naval operations in closer ocean areas. Other nations that might move into this category (in the medium to long term) are Japan and China. Both may eventually move through this level into Rank 2, but not for several decades. On the same time-scale Britain and France may draw closer together in a more united Europe and with the other members form the basis of a Rank 2 navy. If they do not move forward together it is conceivable that the relative decline of British and French capabilities in relation to those of the rest of the world might continue and one, or both, may be reduced to Rank 4.

Rank 4: Medium Regional Force Projection Navy

A Rank 4 navy was defined by Morris as possessing the 'ability to project force into the adjoining ocean basin'.[35] The three Asian members of this group currently are India, Japan, and China. In Europe, Italy, The Netherlands, West Germany, Spain, and even Belgium qualify given their overall maritime capabilities, including land-based aircraft and/or the proven ability to operate as integral parts of NATO's or the WEU's maritime forces at some distance from their own shores. Canada qualifies currently for the same reasons – her ships are often seen in European waters. Her possible purchase of nuclear submarines would, however, have confirmed her status at the top end of Rank 4. The same is true for India which will probably remain a dominant regional power even if China and Japan eventually pursue more global ambitions. In the southern seas, Australia is clearly a Rank 4 navy with a capacity to deploy usable and useful maritime power far into the Pacific. South America has

Three theoretical frameworks

two Rank 4 powers, Brazil and Argentina, each of which currently possesses carriers. Brazil will probably be the first Latin American nation to acquire SSNs.

Rank 5: Adjacent Force Projection Navies

Rank 5 navies have 'some ability to project force well offshore'.[36] They include NATO's Portugal, Greece, and Turkey and the two major non-carrier South American navies, Chile and Peru. Israel, although superficially a territorial defence navy, has often displayed a surprising capacity to strike far from her shores, and South Africa has some capacity to do the same. The two Koreas have navies that are moving into this category, while Taiwan also should not be underestimated with her new submarines and impressive surface fleet. Pakistan and New Zealand both possess potentially long-legged fleets, while the three major Gulf navies – Iran, Iraq, and Saudi Arabia – should soon be in this category if they exploit the equipment they have purchased. None of the above, however, will be capable of carrying out high-level naval operations over oceanic distances.

Rank 6: Offshore Territorial Defence Navies

Rank 6 navies have relatively high levels of capability in defensive (and constabulary) operations up to about 200 miles from their shores. They include the navies of most of the remaining European nations: Norway, Denmark and Sweden, East Germany, Poland, Romania, Bulgaria, and Yugoslavia. Most North African navies are in this category: Algeria, Morocco, Libya, and Egypt. Nigeria is the only West African navy at this relatively high level by African standards, and the mid-level Latin American navies – Cuba, Colombia, Ecuador, and Venezuela – all qualify. In Asia, Bangladesh, Indonesia, Malaysia, and Thailand are clearly in this category, and the Philippines just qualifies. All these navies have the sustainability offered by frigate or large corvette vessels and/or a capable submarine force.

Rank 7: Inshore Territorial Defence Navies

Rank 7 navies have 'primarily inshore territorial defence capabilities'.[37] This is interpreted here to mean navies capable of coastal combat rather than constabulary duties alone. This implies missile-armed, fast-attack craft (sometimes backed up by short-range strike aviation) and sometimes a limited submarine force. These

Conclusion

navies include Albania, Angola, Bahrain, Brunei, Cameroon, Ethiopia, Finland, Gabon, Kenya, Kuwait, North Yemen, Oman, Qatar, Somalia, Singapore, South Yemen, Syria, and Tunisia. Some of these navies, e.g. Kuwait, are trying to acquire more comprehensive capabilities, e.g. MCM vessels.

Rank 8: Constabulary Navies

The Rank 8 category comprises significant navies that are not intended to fight but to act purely in the constabulary role. They do not normally have much beyond guns as armament. The United States, Argentina, Canada, Japan, Norway, India, and Pakistan deploy such forces in addition to their more combatant navy. Mexico's large navy is effectively a maritime police force, as are the Dominican Republic's and Uruguay's rather less impressive forces. Iceland and Ireland have small but highly effective constabulary navies. In Asia, Burma's and Sri Lanka's navies have few capabilities beyond a policing function, while in Africa, Ghana and Tanzania have similarly limited forces. Some of these latter navies may acquire more combatant capabilities and move into Rank 7, but that will depend on the perceived level of threat.

Rank 9: Token Navies

The 'token navy' category covers the rest of the world's navies. These states, the world's smallest and weakest, have some minimal naval capability but this often consists of 'a formal organisational structure and a few coastal craft but little else'.[38] These relatively poor countries cannot aspire to anything but the most limited constabulary functions and it is unlikely they will develop into anything more.

Conclusion

The conclusions to be drawn from this typology of navies seem to be that one might expect the naval balance to change more towards the top than further down the scale. The future hierarchy of the world's navies may well see four or five global navies, USA, USSR, Japan, China, and 'Europe', each with more limited capabilities than the current US Navy. Under these will come the significant regional navies of India, the Americas, and Australia; then will come the lesser African and Asian navies, and the two mavericks of Israel and South Africa; then will come the longer-ranged and the shorter-ranged coast defence navies, the coastguards; and finally the mass of token navies.

Whatever the pecking order, however, sea power will still be more than a mere slogan. It will be a vital factor in the world political order. Countries will have good reason to care about what goes on at sea and they will want, within their means, to have some way of exerting some level of force there. Maritime forces will continue to absorb large amounts of resources, depending on the capacity of the nation to invest them and its perception of the importance of the various uses of the sea, military and civilian, to its overall policy. Certain countries may choose to dismantle their maritime capabilities but others will build them up to compensate. Both seaborne transport and seaborne military power projection will remain of key importance. There will be plenty of scope for the threat or use of force at or from the sea. Sea power, in short, has a sound and secure future.

Notes and references

Chapter 1 Sea power in the modern world

1. A.T. Mahan, *The Influence of Sea Power upon History 1660–1783* (Boston: Little Brown, 1890), p. 1. Page references are from the 1965 Methuen, University Paperbacks facsimile edition.
2. Ibid., p. 26.
3. 'General Council of British Shipping, Statistical Brief, First Quarter 1987', Table 3.
4. *Jane's Fighting Ships 1987–88*, pp. 343, 407.
5. Jean Labayle Couhat (ed.), *Combat Fleets of the World 1986–87* (Annapolis, Md: Naval Institute Press), p. 283.
6. Figures kindly provided by R.S. Woolrych of the British Maritime Charitable Foundation.
7. P.M. Kennedy, *The Rise and Fall of British Naval Mastery* (2nd edition) (London: Macmillan, 1983).
8. H.J. Mackinder, *Britain and the British Seas* (Oxford: Oxford University Press, 1925), p. 358.
9. 'The Geographical Pivot of History' presented to the Society on 25 January 1904, *Geographical Journal*, Vol. 23, No. 4, April 1904, pp. 420–1.
10. Kennedy, op. cit., p. 183.
11. *Geographical Journal*, Vol. 23, No. 4, p. 441.
12. This point was and is often made by Professor Bryan McL. Ranft, late of the Royal Naval College, Greenwich and King's College, London.
13. A.T. Mahan, *The Influence of Sea Power upon the French Revolution and Empire 1793–1812* (Boston: Little Brown, 1892), Vol. 2, p. 118.
14. Sir Julian Corbett, *Some Principles of Maritime Strategy* (London: Longmans, 1911). Quotations come from the new 1988 edition edited by E.J. Grove (Annapolis: US Naval Institute; London: Brassey's, 1988), in this case from p. 15.
15. Lord Esher, *Journals*, 14 August 1914.
16. Admiralty message of 23 November 1918 quoted in D. Vander Vat, *The Grand Scuttle* (Leith: Waterfront, 1986), p. 122.
17. For a stimulating account of the British Empire's victory in 1918 see J. Terraine, *To Win a War* (London: Sidgwick and Jackson, 1978).
18. Most recently in R. Hough's *The Great War at Sea* (Oxford: Oxford University Press, 1983).

19. For example, Kennedy, op. cit., pp. 253-4.
20. See M. Olson, 'The Economics of Wartime Shortage': A History of British Food Supplies in the Napoleonic War and in World Wars I and II (Duke University Press, Durham, NC, 1963). This is an excellent work and deserves greater attention.
21. Kennedy, op. cit., p. 254.
22. See Marder, From the Dreadnought to Scapa Flow, Vol. 4, '1917 the Year of Crisis' (Oxford: Oxford University Press, 1969).
23. S.W. Roskill, The War at Sea, Vol. III, Part II (London: HMSO, 1961), pp. 369-70.
24. See E. Grove and G. Till, 'Adjustments in Anglo-American strategy as a consequence of the Cold War, NATO and nuclear weapons: 1945-67', in Maritime Strategies of the Great Powers: Anglo-American Perceptions (London: Macmillan, 1989).
25. For the playing out of this drama in Britain see E. Grove, Vanguard to Trident, British Naval Policy since World War Two (Annapolis: Naval Institute Press; London: Bodley Head, 1987), pp. 270-7.
26. The Times Atlas of the Oceans (London: Times Books), p. 136.
27. 'Memorandum for the Transport Committee House of Commons Inquiry into the Decline of the U.K. Registered Merchant Fleet', November 1986, British Maritime League, London.
28. F.T. Jane, All the World's Fighting Ships (London: Sampson Low, Marston and Co., 1898), p. 8.
29. F.T. Jane, Fighting Ships, 1905-6 (London: Sampson Low, Marston and Co., 1905), pp. 15-16, 358.
30. Jane's Fighting Ships, 1987-88 (London: Jane's, 1987), p. 105.
31. See The Maritime Strategy, supplement issued with US Naval Institute Proceedings, January 1986, containing article of same name by Admiral J.D. Watkins, the then Chief of Naval Operations. This brochure was the first official widely published statement of The Maritime Strategy. It also contained articles by Secretary of the Navy Lehman and the Commandant General of the Marine Corps. See also 'The 600 Ship Navy and the Maritime Strategy, Hearings Before the Seapower and Strategic and Critical Materials Subcommittee of the Committee on Armed Services, House of Representatives, 99th Congress, First Session, 24 June, 5, 6 and 10 September 1985' (Washington DC: US Government Printing Office, 1986). For an excellent survey of the debate stimulated by the Maritime Strategy see the survey drawn up by one of the Strategy's main drafters, Captain P.M. Swartz USN, Contemporary U.S. Naval Strategy: A Bibliography (Annapolis: US Naval Institute, 1987). An abridged summary appeared in Proceedings February 1987.
32. Notably, N. Friedman, The Post War Naval Revolution (London: Conway, 1986). See also Grove and Till, op. cit.

Chapter 2 Maritime strategy

1. Sir Julian Corbett, Some Principles of Maritime Strategy (London:

Notes and references

Longmans, 1911); and edited by E.J. Grove (Annapolis: US Naval Institute; London: Brasseys, 1988).
2. F.T. Jane, 'Under the white ensign', *Evening Standard*, 15 January 1912.
3. Corbett, op. cit., p. 9.
4. Ibid., p. 94.
5. Ibid., p. 91.
6. S. Turner, 'Missions of the U.S. Navy', *Naval War College Review*, March–April 1974.
7. Ibid.
8. Corbett, op. cit., p. 77.
9. G. Till, *Maritime Strategy and the Nuclear Age* (London: Macmillan, 1982), p. 16.
10. Personal experience, and conversations with my ex-colleague at Dartmouth, Evan Davies.
11. For example, see *Influence of Sea Power Upon History*, p. 24. Mahan could not, however, refrain from falling into the trap of trying to invest the battle itself with special significance. The result is an interesting inversion of logic: 'England was saved at Trafalgar, though the Emperor had then given up his intended invasion.' For Corbett's views, see *The Campaign of Trafalgar* (London: Longmans, 1910).
12. For good accounts of these battles see H.R. Willmott, *The Barrier and the Javelin* (Annapolis: Naval Institute Press, 1983) and W.D. Dickson, *The Battle of the Philippine Sea* (Shepperton: Ian Allan, 1975).
13. S.W.C. Pack, *Night Action off Cape Matapan* (Shepperton: Ian Allan, 1972), p. 19.
14. N. Friedman, *The Postwar Naval Revolution* (London: Conway, 1986).
15. The classic exposition of the Maritime Strategy was *The Maritime Strategy*, supplement to US Naval Institute *Proceedings*, especially the article by Admiral J.D. Watkins 'The Maritime Strategy'. The best explanation, however, is Norman Friedman's *The U.S. Maritime Strategy* (London: Jane's, 1988).
16. Bound lecture notes, Corbett Papers, National Maritime Museum.
17. Corbett, *Strategical Terms and Definitions Used in Lectures on Naval History*, 1906; Corbett Papers. This is reproduced as an appendix in the 1988 edition of *Some Principles of Maritime Strategy*, pp. 324–5.
18. Vice-Admiral Sir Geoffrey Dalton, 'NATO's Maritime Strategy', in E. Ellingsen (ed.) *NATO and U.S. Maritime Strategy* (Oslo: The Norwegian Atlantic Committee, 1987), p. 42.
19. See E. Grove, *Vanguard to Trident, British Naval Policy since World War Two* (Annapolis: Naval Institute Press; London: Bodley Head, 1987), p. 367.
20. Kennedy, op. cit., p. 79.
21. Lt. Cdr. D.W. Waters, *A Study of the Philosophy and Conduct of Maritime War*, Historical Section, Admiralty, revd. ed. 1957, p. 3. A published version of paper appeared in RUSI *Journal*, August 1954.
22. See also J. Winton, *Convoy, The Defence of Sea Trade* (London: Michael Joseph, 1983).
23. Especially in *Sea Power in its Relations to the War in 1812* (New York,

Notes and references

1903). For Mahan's views on convoy see also R.A. Bowling's excellent *The Negative Influence of Mahan on the Protection of Shipping in Wartime: The Convoy Controversy in the Twentieth Century*, Ph.D. thesis presented to the Graduate School University of Maine at Orono, May 1980. For a more easily accessible and recent variant of Bowling's well argued views see his 'Keeping Open the Sea Lanes', US Naval Institute *Proceedings*, December 1985, pp. 92-8.

24. Waters, op. cit.
25. Marder, *From the Dreadnought to Scapa Flow*, Vol. 4; '1917 the Year of Crisis'.
26. 'Defence of Trade', *Naval War Manual 1925*, p. 35; Public Record Office reference ADM186/66.
27. Waters, op. cit., p. 32.
28. Ibid., p. 36.
29. Watkins, op. cit., p. 11.
30. Waters, op. cit., p. 37.
31. Ibid., p. 39.
32. Quoted, ibid., p. 16.
33. Ibid.
34. Letter from Cdr. Waters to Cdr. M. Ranken, 3 October 1981.
35. D.W. Waters, 'Seamen, Scientists, Historians and Strategy', Presidential Address, 1978, *The British Journal of the History of Science*, Vol. 13, No. 45, 1980 (footnote 29).
36. Ibid., pp. 207-9; I have made one or two small alterations of expression.
37. This formula was produced by Lord Blackett in 1943. Blackett was perhaps the most intellectually distinguished product of the Royal Naval Colleges at Osborne and Dartmouth during their most dynamic years before 1914.
38. Ibid., p. 210.
39. E.J. Grove, 'The reluctant partner: the United States and the introduction and extension of convoy 1917-18', in S. Palmer and G. Williams (eds), *Charted and Uncharted Waters* (Proceedings of a Conference on the Study of British Maritime History; National Maritime Museum and Queen Mary College, London, 1981).
40. See Corbett, op. cit., pp. 214-20, for a sympathetic account of Torrington's activities.
41. Retired Hungarian general, in conversation with the author.
42. For the 'Jeune Ecole' and German thought, see T. Ropp, 'Continental doctrines of sea power', in E.M. Earle (ed.), *Makers of Modern Strategy* (Princeton University Press, 1941), chap 18. For a more modern assessment, see G. Till (1982) op. cit., pp. 34-8, 49-53.
43. For Soviet thinking, see B. Ranft and G. Till, *The Sea in Soviet Strategy* (London: Macmillan, 1983).
44. For a good discussion of this matter see G. Till, *Modern Sea Power* (London: Brasseys, 1987), pp. 110-22. See also Chapter 4.
45. Andrew Lambert, 'Maritime Strategy against Russia: An Historical Perspective', unpublished ms.
46. Ibid.
47. Ibid.

Notes and references

48. Ibid.
49. Ibid.
50. Corbett, op. cit., title of Part II, chap. II.
51. Ibid., p. 107.
52. Ibid., p. 111.
53. Ibid., pp. 123–7.
54. W.H. Garzke and R.O. Dulin, *Battleships: Axis and Neutral Battleships in World War II* (Annapolis: Naval Institute Press, 1985).
55. For the best discussion of Fisher and the battlecruiser see J. Sumida, *In Defence of Naval Supremacy: Technology, Finance and British Naval Policy* (London: Allen and Unwin, 1989), chaps 2, 3 and 5.
56. Fisher, *Records* (London, 1919), pp. 212–3.
57. Admiral Sir John McGeoch, 'Command of the sea in the seventies', (Waverley paper, Edinburgh University, 1972).

Chapter 3 Economic uses of the sea

1. UN General Assembly Fortieth Session, Report to the Secretary General, 'Study on the Naval Arms Race'.
2. British Maritime League – Memorandum for the Transport Committee, House of Commons, *Inquiry into the Decline in the U.K. Registered Merchant Fleet*, November 1986, p. 7.
3. Ibid., and the GCBS Statistical Brief, 1st Quarter 1987, p. 8.
4. GCBS Brief.
5. BP Statistical Review of World Energy, June 1987, p. 19.
6. Ibid., and *Times Atlas of the Oceans* (London: Time Books, 1983), pp. 146–7, 150–1.
7. *Why the Ships Went: A Fully Quantified Account of the Clauses and Decline of the U.K. Registered Merchant Fleet* (British Maritime Charitable Foundation and Lloyds of London Press, 1986), p. 52. For tanker and gas carrier development see Carlett, *The Ship: The Revolution in Merchant Shipping 1950–80* (London: HMSO, 1981), pp. 24–30.
8. *Why the Ships Went*, p. 25; *Times Atlas*, op. cit., p. 138.
9. Carlett, op. cit., pp. 39–43; *Times Atlas*, op.cit., p. 138.
10. GCBS Brief: *Times Atlas*, op. cit., pp. 136–9.
11. Ibid.
12. Carlett, op. cit., p. 9 and note.
13. Ibid., p. 10.
14. *Times Atlas*, op. cit., p. 139.
15. Ibid.
16. Ibid., p. 13.
17. Ibid.
18. Ibid.
19. A.J. Ambrose, 'The supercarriers', *Jane's Merchant Shipping Review*, 3rd Year of Issue (London: Jane's, 1985).
20. Carlett, op. cit., p. 21; for the new types of vessels; see also A.J. Ambrose, 'The big ship market – tankers and bulk carriers', *Jane's*

Notes and references

Merchant Shipping Review, First Year of Issue, (London: Jane's, 1983), pp. 32–41; and 'British yards set to exploit the future', *Marine Engineer's Review Supplement*, October 1986, p. 5.
21. Carlett, op. cit., pp. 17–20.
22. A.J. Ambrose, 'U.S. lines big twelve', *Jane's Merchant Shipping Review*, 3rd Year of Issue (London: Jane's, 1985).
23. A.J. Ambrose, 'The traumas of the TEU transporters', *Jane's Merchant Shipping Review*, 3rd year of Issue (London: Jane's, 1985), p. 163. Up-to-date lists of container facilities are also contained in the annual editions of *Jane's Freight Containers*.
24. N. Polmar, *Guide to the Soviet Navy*, 4th Edition (Annapolis: Naval Institute Press, 1986), p. 463.
25. M. Enquist, 'The reefer trades in the mid 1980s', *Jane's Merchant Shipping Review*, 3rd Year of Issue (London: Jane's, 1985).
26. Carlett, op. cit., pp. 44–5 is a brief history of this type of vessel.
27. A.J. Ambrose, 'Development in BASH (barge aboard ship) vessels', *Jane's Merchant Shipping Review*, 2nd Year of Issue (London: Jane's, 1984), pp. 132–8; Polmar, op. cit., pp. 369–73, 483.
28. G.W. Detlefsen, 'The German Coaster Fleet', *Jane's Merchant Shipping Review*, 2nd Year of Issue, pp. 107–12.
29. 'British Yards' (see note 20), ibid.
30. Carlett, op. cit., pp. 33–8.
31. UN Conference on Trade and Development, 'Review of Maritime Transport, 1983' (TD/B/C.4/266), p. 11.
32. GCBS, Statistical Brief No. 16, pp. 3–4.
33. Ibid.
34. *Times Atlas*, op. cit., p. 141.
35. Ibid., tables on pp. 141–2.
36. Figures kindly provided by R. Woolrych, the British Maritime Charitable Foundation.
37. *Why the Ships Went*, pp. 13–15; Polmar, op. cit., pp. 479–80. For a full survey of American merchant shipping policies see C.H. Whitehurst, Jr., *The US Merchant Marine in Search of an Enduring Maritime Policy* (Annapolis: Naval Institute Press, 1983).
38. *The Naval Arms Race*, p. 7 (see note 1).
39. *Britannica Book of the Year* 1987, p. 121.
40. *The Naval Arms Race*, p. 7.
41. *Times Atlas*, op. cit., p. 97.
42. Ibid.
43. *Britannica Book of the Year* 1987, p. 121.
44. Ibid.
45. *Arms Race*, p. 7; *Britannica Book of the Year* 1986, p. 136; 1987, pp. 121–2.
46. *Britannica Book of the Year* 1986, 'Iceland', p. 499; 1987, pp. 198, 454.
47. Ibid., see also 1986, pp. 137, 223.
48. *Times Atlas*, op. cit., p. 104.
49. *Britannica Book of the Year* 1988, p. 723.
50. *Times Atlas*, op. cit., p. 104.

Notes and references

51. Ibid., pp. 110–17, *Arms Race*, pp. 8–9; also author's discussions with W. Ostreng at the Nansen Institute, Oslo, September 1986.

Chapter 4 The military medium

1. A.T. Mahan, *The Influence of Sea Power upon History 1660–1783* (Boston: Little Brown, 1890), p. 25.
2. J. Sokolsky, 'Seapower in the nuclear age: Nato as a maritime alliance', thesis presented to the Dept. of Government, Harvard University, Cambridge, Mass., August 1984, p. 278.
3. Ibid.
4. Organization of the Joint Chiefs of Staff: *Military Posture Fiscal Year 1987*, p. 67.
5. Sokolsky, op. cit., p. 268.
6. Mahan, op. cit., p. 28.
7. *The Military Balance 1987* (London: IISS, 1988).
8. S. Gorshkov, *Navies in War and Peace* published as *Red Star Rising at Sea* (Annapolis: Naval Institute Press, 1974) and *The Seapower of the State* (Oxford: Pergamon, 1979).
9. UN General Assembly Fortieth Session, Report to the Secretary General, 'Study on the Naval Arms Race', p. 1.
10. T.A. Gibson, 'Gallipoli: 1915', in Merrill L. Bartlett (ed.) *Assault from the Sea* (Annapolis: Naval Institute Press, 1983), p. 152.
11. Ibid.
12. Colonel Robert D. Heinl Jr., USMC, 'The U.S. Marine Corps: author of modern amphibious warfare', in Bartlett, op. cit., p. 187.
13. H.G. Von Lehmann, 'Japanese landing operations in World War II', in Bartlett, op. cit., pp. 195–201.
14. K.J. Clifford, *Amphibious Warfare Development in Britain and America from 1920–1940* (Laurens NY: Edgewood Inc., 1983), p. 249.
15. Vice-Admiral B.B. Schofield, *Operation Neptune* (Shepperton: Ian Allan, 1974), pp. 54–5.
16. Ibid., pp. 54–5, 120.
17. S. Roskill, *The Strategy of Sea Power* (London: Collins, 1962), p. 240.
18. Ibid., p. 243.
19. M. Binkin and J. Record, *Where Does the Marine Corps Go from Here* (Washington: The Brookings Institution, 1976), p. 14.
20. Ibid., p. 6; see also R.J. Heinl Jr., 'Inchon, 1950', in Bartlett, op. cit., this also covers briefly post-war opposition to the amphibious mission.
21. See E. Grove, *Vanguard to Trident, British Naval Policy since World War Two* (Annapolis: Naval Institute Press; London: Bodley Head, 1987), especially chaps 5, 7, 8 and 10.
22. Captain R.S. Moore USMC, 'Blitzkrieg from the sea: maneuver, warfare and amphibious operations', *Naval War College Review*, November–December 1983.
23. Discussion of the underwater environment is based on D.C. Daniel, *Antisubmarine Warfare and Superpower Strategic Stability* (London: Macmillan for IISS, 1986), chap 2. See also T. Stefanick, *Strategic Antisubmarine Warfare and Naval Strategy* (Lexington, Mass.: Lexington

Notes and references

Books, 1987); Rear-Admiral J.R. Hill, *Anti-Submarine Warfare*, pp. 37–43; and R. Compton Hall, *Submarine Versus Submarine* (Newton Abbot: David and Charles, 1988), p. 40. The UEP mine is mentioned on p. 63. *The Times Atlas of the Oceans*, pp. 50–1, is also a useful guide to the underwater environment.

24. Daniel, op. cit., pp. 70–84 gives a good account of non-acoustic submarine indicators; see also Stefanik, op. cit., Appendix 3, pp. 181–215.
25. Grove, op. cit., p. 291; D. Wettern, *The Decline of British Sea Power* (London: Jane's, 1982), p. 292; *Navy International*, July 1967, p. 244.
26. Grove, op. cit., pp. 358–9.
27. For US naval submarine intelligence activities see J.T. Rachelson and D. Ball, *The Ties That Bind* (Boston: Allen and Unwin, 1985), pp. 223–4.
28. *Jane's Fighting Ships 1987-8*, pp. 617–20.
29. D. Pearson, *KAL007, The Cover Up* (New York: Summit Books, 1987), pp. 58–9.
30. *The Naval Arms Race*, Dept. for Disarmament Affairs, Report of the Secretary General (New York: United Nations, 1986), p. 66.
31. Ibid., p. 67.
32. *Vanguard to Trident*, pp. 366–7.
33. L. Martin, *The Sea in Modern Strategy* (London: Chatto and Windus, 1965), p. 87.
34. Sokolsky, op. cit., p. 138.
35. CICLANTFLT report to OPLAN 2300, quoted in Sokolsky, op. cit., p. 141.
36. Sokolsky, op. cit., p. 142.
37. Sir Julian Corbett, *Some Principles of Maritime Strategy* (London: Longmans, 1911); and edited by E.J. Grove (Annapolis: US Naval Institute; London: Brasseys, 1988), p. 12.

Chapter 5 Weapon systems

1. N. Friedman, *U.S. Naval Weapons* (London: Conway, 1983), p. 24.
2. For submarine operations in the Falklands War see E. Grove, *Vanguard to Trident, British Naval Policy since World War Two* (Annapolis: Naval Institute Press; London: Bodley Head, 1987), pp. 357–81 and David Brown, *The Royal Navy in the Falklands War* (London: Leo Cooper, 1987).
3. R. Scheina, 'Where were those Argentine subs?', US Naval Institute *Proceedings*, March 1984.
4. 'Plans unveiled for ASUW torpedo', *Jane's Defence Weekly*, 5 April 1986, p. 600.
5. See 'underwater warfare' section, *Jane's Weapons Systems 1988-9* (London: Jane's, 1988).
6. Donald C. Daniel, *Anti-Submarine Warfare and Superpower Strategic Stability* (London: Macmillan, 1986), p. 45. For Soviet torpedoes see also N. Polmar, *Guide to the Soviet Navy* (Annapolis: Naval Institute Press, 1986), p. 435.
7. For decoys see R.J.L. Dicker, 'Decoying the homing torpedo', *International Defense Review*, 2/85; for anti-torpedo torpedoes see J. Hutsebaut,

Notes and references

'Still no cure for the torpedo', United States Naval Institute, *Proceedings*, October 1985, p. 147.
8. See diagrams in *Aviation Week and Space Technology*, 2 February 1976 pp. 12, 15.
9. Polmar, op. cit., p. 431.
10. E. Briffa and G. Williams gave a good description of Exocet's *modus operandi* in their television programme 'In the Wake of HMS Sheffield' broadcast twice on BBC 2 in 1986. For the problems of long-range engagement, see N. Friedman op. cit., p. 231.
11. Briffa and Williams, op. cit; see also *Jane's Weapons Systems 1988-89* (London: Jane's, 1988).
12. For ISAR see Friedman, *The U.S. Maritime Strategy*, p. 148.
13. M. McCGwire, *Military Objectives in Soviet Foreign Policy* (Washington: Brookings Institution, 1987), p. 411; *Aviation Week and Space Technology* published a mission profile of SS-NX-13 on p. 14 of the 2 February 1976 issue.
14. For novel British weapons see L. Marriott, 'Flights of fancy', *Warship World*, Spring 1987, pp. 10-11; N. Friedman, op. cit., pp. 118-19 for Petrel.
15. Wallop Systems Ltd., *Barricade Naval Decoy Systems*; Marconi Defence Systems, *Siren and Towed Decoy*.
16. Brown, op. cit., pp. 194-5. Even retarded bombs need a minimum height to arm and allow the aircraft to get clear.
17. For analyses of mine warfare and mine countermeasures see C. Rouarch, 'The naval mine: as effective a weapon as ever', *International Defense Review*, September 1984; M. Vego, 'Soviet mines and mine warfare', *Navy International*, July 1986; *Navy International*, December 1986; H.J. Bangen, 'The impact of modern technology on the performance of the sea mine', *Naval Forces*, Vol. VI, No. I, 1985; Captain C.C.M. de Nooijer, 'Countering the mine threat', *Naval Forces*, Vol VI, No. IV, 1985; S.E. Fawson, 'Mine warfare – NATO versus the Warsaw Pact', *Naval Forces*, Vol. VII, No. V, 1986; R.J.L. Dicker, 'Mine warfare now and in the 1990s', *International Defense Review*, No. 3, 1986; for a historical perspective see G.K. Hartman, *Weapons that Wait*, (Annapolis: Naval Institute Press, 1979). For ARMS see *Advance Remote Minehunting System* (Plessey brochure, 1987); for SAM see, *MCM News* (Malmo: Arbetsgruppen, 1988).
18. Captain B.R. Longwirth, 'The lightweight torpedo' *Naval Forces*, Vol. VI, No. VI, 1985, p. 94-5.
19. Ibid., p. 21.
20. House of Commons, Twenty-eighth Report from the Committee of Public Accounts, Session 1984-85, *The Torpedo Programme* HC391; E.R. Hooton, 'The British torpedo: on track at last', *Military Technology*, Vol. IX, Issue 7, 1986.
21. Hooton, op. cit., pp. 92-4; 'Anti-submarine warfare – the torpedo', *Navy International*, May 1987, pp. 274-83.
22. Ibid.
23. See *Jane's Weapons Systems* for performance data on the Mk 46.

Notes and references

24. Longwirth, op. cit.
25. 'Anti-submarine warfare – the torpedo', *Navy International*, May 1987.
26. Polmar, op. cit., pp. 435–6.
27. For British decoys see R.J.L. Dicker, 'Avoiding the homing torpedo', op. cit.; for US equipment see *International Defense Review*, June 1987, p. 811. For double hulls and torpedoes see N. Friedman, 'A survey of western ASW in 1985', *International Defense Review*, No. 10, 1985, p. 1597.
28. See *Jane's Weapon Systems* for details of these systems; and N. Polmar, *Ships and Aircraft of the U.S. Fleet* (Annapolis: US Naval Institute, 1987) p. 474 for Sea Lance.
29. Ibid., see also Compton Hall, *Submarine Versus Submarine*, pp. 61–3 and the brochure *Crusader Advanced Sea Mine* (British Aerospace, 1988).
30. Brown, op. cit., pp. 102, 137.
31. See *Jane's Weapon Systems* for a useful annual survey of these devices.
32. 'A new lightweight ASW system against submarine intruders in shallow coastal waters', paper presented to the Royal Institution of Naval Architects Conference, Anti-Submarine Warfare, May 1987.
33. Polmar, op. cit., pp. 409–11.
34. N. Friedman, 'A survey of western ASW', *International Defense Review*, No. 10, 1985; also D. Wettern, 'Active sonars and warship design', *Navy International*, May 1987.
35. Friedman, op. cit., and *The Postwar Naval Revolution* (London: Conway, 1986) for a discussion of this change. Friedman summarizes early towed array developments in 'The evolution of towed array sonar systems', *Naval Forces*, Vol. IV, No. V, 1983. See also his *The US Maritime Strategy* (London: Jane's, 1988) for more useful information on SOSUS and its operational exploitation: FDS is mentioned on p. 129.
36. N. Friedman, 'Sonar and sensing in the far deep', *NATO's Sixteen Nations*, July 1987, p. 79.
37. P. Sabathe, 'ASW sonars for surface ships', *International Defense Review*, No. 10, 1985, p. 1603. For the combined 'active towed array sonar' see *ATAS* (British Aerospace brochure, 1988).
38. Friedman (1987), op. cit.; presentation by Commodore R. Cobbold, 'Future Maritime Means and Needs', Royal United Services Institute, 16 November 1988, quotation from C.C.F. Adcock, Plessey Naval Systems, *Some Aspects of Bistatic Sonar*, Undersea Defence Technology Conference/Exhibition, October, 1988.
39. Friedman (1987), op. cit., and *HELRAS* (British Aerospace brochure).
40. E.R. Hooton, 'Sonobuoys', *Naval Forces*, Vol. V, No. VI, 1984; and G.S. Sundaram, 'Sonobuoys and dipping sonars: remote sensors for airborne ASW' *International Defense Review*, March 1983; M. Witt, 'Sonobuoys', *Navy International*, May 1987.
41. For an excellent description of ASW in all its facets see Rear-Admiral J.R. Hill, *Anti-Submarine Warfare* (Shepperton: Ian Allan, 1984).
42. Friedman (1983), op. cit., pp. 24, 168; Brown, op. cit., p. 256.
43. For 'Aegis' see Captain Joseph L. McClane and Commander J.L. McClane, 'The Ticonderoga story: Aegis works', US Naval Institute

Notes and references

Proceedings, May 1985; also Friedman (1983), op. cit., pp. 158-9; N. Polmar, *The Ships and Aircraft of the U.S. Fleet*, pp. 506-7, 479, 112-16, 139-140.

44. *Jane's Defence Weekly*, 19 April 1986, pp. 702-3; 3 October 1987, p. 740; 17 October 1987, p. 864; for Aster see *Maritime Defence*, October 1988. pp. 381-2.
45. Lt. Cdr. W.A. Weronkoo in US Naval Institute *Proceedings*, November 1984, p. 183.
46. Polmar, op, cit., pp. 424, 448 for details of SA-N-6 missile and radar. For Baku see *Jane's Defence Weekly*, October 1988, p. 893.
47. Brown, op. cit., p. 222; for Sea Dart's successes see *Falklands, the Air War* (British Aviation Research Group, 1986); and E.J. Grove, *Vanguard to Trident, British Naval Policy since World War Two* (Annapolis: Naval Institute Press; London: Bodley Head, 1987).
48. Friedman (1983), op. cit., pp. 151-2.
49. As in *Coventry's* loss; Brown, op. cit.
50. See *Jane's Weapon Systems* for Sea Wolf details and variants.
51. Friedman (1983), op. cit., p. 182.
52. J. Ethell and A. Price emphasize this point in their *Air War South Atlantic*, (London: Sidgwick and Jackson, 1983).
53. See map on p. 13 of *The Maritime Strategy*, supplement to US Naval Institute *Proceedings*, January 1986. See also N. Friedman, *Carrier Air Power*, (London: Conway, 1981), chap 7.
54. *The Maritime Strategy*, op. cit. For an excellent overview of the whole air defence problem see J.R. Hill, *Air Defence At Sea* (Shepperton: Ian Allan, 1988).
55. For a Soviet example see US Department of Defense, *Soviet Military Power* (Washington: Government Printing Office, 1987), p. 113.
56. For example see S. Gorshkov, *The Seapower of the State* (Oxford: Penguin, 1979), chap IV.
57. For the potential of battleships for shore bombardment and their limitations see E.J. Grove, 'U.S. Navy Battleship Reactivation - A Commentary', *Naval Forces*, Vol. VIII, No. III, 1987, pp. 85-6.
58. Friedman (1983), op. cit., p. 22; the new British 155 mm naval gun was displayed by Royal Ordnance at the Royal Naval Equipment Exhibition in 1987.
59. R.E. Gottemoeller, 'The modern sea launched cruise missile', *Naval Forces*, Vol. VIII, No. IV, 1987 is a good recent survey of the cruise missile scene.
60. Mike MccGwire, *Military Objectives in Soviet Foreign Policy* (Washington: The Brookings Institution, 1987), pp. 501-2.
61. J. Sumida, *In Defence of Naval Supremacy: Technology, Finance and British Naval Policy* (London: Allen and Unwin, 1989), Epilogue and Conclusions.
62. For an excellent explanation of these developments see Captain F.D. Stanley, 'Action Information Organisation', *Navy International*, December 1985; and N. Friedman, *US Destroyers* (Annapolis: Naval Institute Press, 1982), especially chap. 8.
63. Admiral Sir Arthur Hezlet, *The Electron and Sea Power* (London: Peter Davies, 1975), p. 284.

Notes and references

64. See editorial in *Warship World*, Vol. 1, No. 10, Spring 1987.
65. Stanley, op. cit., p. 753.
66. See 'Command and Control' survey, *Navy International*, January 1987.
67. Ibid., p. 137.
68. Commodore R. Cobbold to Royal United Services Institute, 16 November 1988.
69. Stanley, op. cit., p. 754.
70. For the range of US naval nuclear weapons see Friedman (1983), op. cit., pp. 278–80.
71. R.B. Berman and J.C. Baker, *Soviet Strategic Forces, Requirements and Responses*, (Washington: The Brookings Institution), pp. 54–5.
72. D. Campbell, 'Too few bombs to go round', *New Statesman*, 29 November 1985, describes the British position.
73. Gottemoeller, op. cit., for discussion of TLAM/N and the difficulties it poses.
74. P. Langereux, 'France's ASMP nuclear cruise missile operational', *Military Technology*, vol. IX, Issue 7.
75. As the 1985 Defence White Paper (Cmnd. 9430-1) put it: 'Free-fall nuclear bombs can also be delivered by the Sea Harriers of the squadrons in service with the Royal Navy's "Invincible" class or aircraft carriers.'
76. This targeting policy was revealed by J. Barry in the television programme 'Britain's bomb – the secret story' first broadcast 6 April 1986, Channel 4.
77. For good descriptions of Polaris and Trident see Friedman (1983), op. cit., pp. 220–5.

Chapter 6 Platforms

The main reference for this chapter is *Jane's Fighting Ships 1987–88*, and *Combat Fleets of the World 1988–9*.

1. See Chapter 3.
2. For a set of novel ideas on ships of the future see R. Schonknecht, J. Lusch, M. Schelzel and H. Obenaus, *Ships and Shipping of Tomorrow* (Berlin: Verlag Technik; Hounslow: McGregor Publications, 1983).
3. Ibid., pp. 120, 122.
4. J.R. Hill, *Maritime Strategy for Medium Powers* (London: Croom Helm, 1986), pp. 158–64.
5. H. Moineville, *Naval Warfare Today and Tomorrow* (Oxford: Basil Blackwell, 1983), p. 63.
6. For a survey of offshore patrol vessels see *International Defense Review*, April 1985, pp. 483–509.
7. For a good survey see C.E.R. Borgenstarn, 'Fast attack craft', *Naval Forces*, Vol. VII, No. II, 1986, pp. 69–77 and Supplement 'Fast naval combat craft', *International Defense Review*, January 1985.
8. *Aviation Week and Space Technology*, 31 March 1986.
9. For the High/Low mix concept see N. Friedman, 'Elmo Russell Zumwalt Jr.' in R.W. Love Jr., *The Chiefs of Naval Operations* (Annapolis: Naval Institute Press, 1980), pp. 371–2.

Notes and references

10. See E.J. Grove, *Vanguard to Trident, British Naval Policy since World War Two* (Annapolis: Naval Institute Press; London: Bodley Head, 1987), pp. 204, 217–18.
11. A.H. Lind, 'The re-emergence of the corvette', *Naval Forces*, Vol. VI, No. V, pp. 54–65.
12. N. Friedman, *US Cruisers* (Annapolis: Naval Institute Press, 1984), pp. 419–22.
13. For an analysis of battleship re-activation see E.J. Grove, 'U.S. Navy Battleship Reactivation – A Commentary', *Naval Forces*, vol. VII, No. III, 1987, pp. 84–90.
14. Grove, *Vanguard to Trident*, pp. 343–4.
15. For an excellent and authoritative description of the Type 23's evolution, and a survey of the Royal Navy's procurement practices, see Sir Lindsay Bryson, 'The procurement of a warship', *The Naval Architect*, January 1985.
16. For a debate over the best role for these ships and a good indication of how uneasily such sea control vessels lie in an essentially power projection navy see US Naval Institute *Proceedings*, June 1983.
17. For a profile of the 'Leander' see C.J. Meyer, *Leander Class*, (Shepperton: Ian Allan, 1984).
18. *Modernised Leander Frigate*, Vosper Thorneycroft sales brochure.
19. J.W. Kehoe and K.S. Brower, 'Small combatants – the operators choices', in *Small Warships*, supplement to *International Defense Review*, May 1987; for Greek policy see *Jane's Defence Weekly*, 27 February 1988, p. 335.
20. R.J.L. Dicker, 'The frigate and corvette export market: still in the doldrums', *International Defense Review*, April 1986, p. 449.
21. Ibid., p. 450.
22. L.J. Lamb, 'The Taiwan navy: building to a regional role?', *International Defense Review*, November 1985, pp. 1745–52.
23. Lind, op. cit., pp. 54–64 gives a useful account of modern corvettes; the FS1500s are covered in R.J.L. Dicker, 'HDWs FS1500: the bargain basement frigate', *International Defense Review*, November 1984.
24. Dicker (1986), op. cit., p. 456.
25. Ibid.
26. Vice Admiral Joseph Metcalf III. 'Revolution at Sea, US Naval Institute *Proceedings*, January 1988, p. 37. See also T.M. Keithly, 'Tomorrow's surface forces', US Naval Institute *Proceedings*, December 1988, pp. 50–61.
27. For pro 'short fat' ship arguments see D.L. Giles, 'Want of a frigate', *Naval Forces*, Vol. V, No. II, 1984; and N. Ling, 'The S90 argument', *Jane's Defence Weekly*, 11 October 1986. For the case against see A. Preston, 'Warship controversy', *Jane's Defence Weekly*, 12 July 1986, and N. Friedman, 'S90: A U.S. Viewpoint', *Jane's Defence Weekly*, 11 October 1986. For the author's views at more length see 'Short and fat or long and thin', US Naval Institute *Proceedings*, March 1987. For the quotation see *Report of the Inquiry into Hull Forms for Warships*. (London: HMSO, 1988), p. 11.

Notes and references

28. See Michael J. Eames, 'Future naval surface ships', and Clark Graham, 'The modern monohull', in *Modern Ships and Craft*, a special edition of the *Naval Engineer's Journal*, Vol. 97, No. 2, February 1985.
29. Ibid., p. 56.
30. For catamarans in general see T.F. Carreyette, 'Recent developments in advanced hull types for naval ships', *Naval Forces*, Vol VIII, No. I, 1987. For the specific types see *International Defense Review*, February 1987, pp. 199–201 and *Maritime Defence*, October 1988, p. 377.
31. J. Stebbins, 'Sea Knife arriving', US Naval Institute *Proceedings*, February 1985, pp 113–18.
32. Eames, op. cit., p. 61. See also R.L. Johnston, 'Hydrofoils' in the same work. Carrayette, op. cit., also contains a useful summary of the characteristics of hydrofoils.
33. D. Lavis, 'Air cushion craft', in *Modern Ships and Craft*.
34. British Hovercraft brochure: *BH7 mk20 Multi-Role Amphibious Hovercraft*.
35. E.A. Butler, 'The surface effect ship', *Modern Ships and Craft*, p. 200. For SES see also Carrayette, op. cit.
36. Butler, op. cit., p. 200.
37. R.D. Watkins and P.D. Plumb, 'SEALS and Cigarettes', US Naval Institute *Proceedings*, April 1987, pp. 103–4.
38. Butler, op. cit., p. 209.
39. Ibid., p. 211: this paragraph is based on Butler's excellent analysis. For supplementary discussion from a Norwegian perspective see S.G. Oystein Ronning, 'Needs within the Norwegian Navy/NATO towards the year 2000', *Maritime Defence*, May 1988.
40. Butler, op. cit., p. 211.
41. J.L. Gore, 'SWATH Ships', *Modern Ships and Craft*, p. 94.
42. For an effective critique of SWATH see W.C. Barnes, 'SWATH: advanced technology or mythology', US Naval Institute *Proceedings*, September 1986, pp. 119–21.
43. Gore, op. cit. For a rebuttal of Barnes see A.C. McClure and R.D. Gaul, 'Let SWATH be SWATH', US Naval Institute *Proceedings*, April 1987, pp. 106–8. For propulsion improvements see Carreyette, op. cit., p. 77.
44. Gore, op. cit. For a practical SWATH ASW frigate see G.V. Betts, L.D. Ferroro, S.R. Grzeskoviak, N.A. Macdonald and P.J. Bartlett, 'Design Study for an Anti-Submarine Warfare SWATH Frigate', paper presented to the R.I.N.A. 'Warship 87' Symposium, May 1987.
45. S.E. Veazey, 'New shape in ships', US Naval Institute *Proceedings*, p. 40.
46. Amended from an original illustrating Eames, op. cit., p. 59.
47. CAP and YARD, *Project Capability*, 1987.
48. 'Naval propulsion', *International Defense Review*, 8/1985, pp. 1273–94 is a good survey of propulsion matters. See also D.J. Gates, *Surface Warships* (London: Brasseys, 1987), chap. 5, and Keithly, op. cit., p. 57.
49. Babcock, GEC and Weirs brochure, 'Modern steam propulsion for warships', Section 1, p. 11.
50. John Lehman, *Aircraft Carriers: The Real Choices*, Washington Papers,

Notes and references

Vol. VI, No. 5, Georgetown University, Washington DC, (Beverley Hills and London: Sage Publications, 1978).
51. Ibid., pp. 39-40.
52. Ibid., p. 40.
53. Private information from Soviet sources. See also *Jane's Defence Weekly* 19 April 1986, p. 702 and 27 August 1988, p. 388; and 'New Soviet carrier more like a Kiev than a Nimitz', *International Defense Review*, March 1987. Mike MccGwire, *Military Objectives in Soviet Foreign Policy* (Washington: 1987, The Brookings Institution), pp. 441-2, who argues that changing the design from a 'pure' carrier to such a 'universal ship' allowed the ship to be transferred to a Black Sea yard from Severodvinsk in the north.
54. For a description of *Argus* see Harland and Wolffe's brochure, *RFA Argus: Aviation Training Ship*. See also D. Wettern, 'First Report: RFA Argus', in *Warship World*, Vol. I, No. 7, Summer 1986, and K. Walsh and N.A.J. Wells, 'RFA Argus: The conversion of a Ro-Ro container vessel to an aviation training ship', paper presented to RINA Conference, *Merchant Ships to War*, 1987.
55. N. Polmar, *Ships and Aircraft of the US Fleet* (Annapolis: Naval Institute Press, 1984), pp. 80, 101. For a critique of these ships see Lehman, op. cit., pp. 59-68.
56. See D. Brown, 'The MILCOM vessel – a philosophical view', and N.P. Simpson, 'Merchant ships to warships', papers prepared for the RINA Conference *Merchant Ships to War*, 19 October 1987.
57. Brochure, Sea Containers Ltd: *Aviation Support Ship: The Low Cost Warship*.
58. P.D. Pound, 'Naval auxiliaries: a multi-role naval capability from a merchant hull', paper presented to RINA Conference, *Merchant Ships to War*, 1987. Also brochure, British Shipbuilders, *MRNA*.
59. For an interesting discussion of protecting a modern warship see J.R.L. Dicker. 'Survivability and the modular frigate – new developments from Blohm and Voss', *International Defense Review*, July 1986, pp. 961-4. See also E. Rohkamm, 'Are armoured frigates feasible?', *Naval Forces*, Vol. VI, No. II, 1985.
60. W.T.T. Pakenham, 'The command and control of submarine operations', *Naval Forces*, Vol. VI, No. II, 1985; see also R. Compton Hall, *Submarine Versus Submarine: the tactics and technology of underwater warfare* (Newton Abbott: David & Charles, 1988), pp. 47-50.
61. For the basics of blue-green laser technology see J. Starkey, 'The renaissance in submarine communications', *Military Electronics*, April, 1981. For a useful update and the figures on ELF data see *Jane's Defence Weekly*, 14 November 1987, p. 1097. See also N. Friedman in the Naval Review issue of United States Naval Institute *Proceedings*, May 1987, p. 97.
62. For modern submarine design see R.J. Daniel, 'Submarine design', *Naval Forces*, Vol. VIII, No. II, 1987 and N. Friedman, 'The U.S. next generation attack submarine', *International Defense Review*, March 1984. J.E. Moore and R. Compton-Hall, *Submarine Warfare Today and Tomorrow*

Notes and references

(London: M. Joseph, 1986) gives a stimulating and idiosyncratic survey of SSN developments. For Soviet improvements see also J.E. Moore, 'Soviet submarine technology forges ahead', *Jane's Naval Review*, 6th Year of Issue, pp. 84–8 and 'Soviets score silent success in undersea race with US', *Washington Post*, 17 July 1987.

63. For pump jet references see *Combat Fleets of the World, 1986-87* (Annapolis: Naval Institute Press), pp. 185, 618.
64. *The 600 Ship Navy and the Maritime Strategy*, Hearing Before the Strategic and Critical Materials Sub-Committee of the Committee on Armed Services, House of Representatives, Ninety Ninth Congress, First Session, p. 163.
65. For the rebuttal see ibid., for the assertion see Moore and Compton Hall, op. cit., p. 153, and Moore, op. cit. See also Compton Hall, op. cit., p. 56.
66. See N. Friedman in US Naval Institute *Proceedings*, August 1987, p. 133: also *Jane's Defence Weekly*, 17 October 1987, p. 853, 30 January 1988, and 13 February 1988.
67. *Jane's Defence Weekly*, 19 December 1987, pp. 1399, 1445; 30 January 1988, p. 152.
68. N. Friedman, 'Conventional submarine developments', *Naval Forces*, Vol. VI, No. VI, 1985. For good surveys see 'Conventional submarines and their combat systems', *International Defense Review*, July 1987, pp. 927–34 and the March 1988 edition of *Maritime Defence*. The Stirling system was covered well in a supplement to *Naval Forces*, Vol VIII, No. VI, 1987.
69. *Combat Fleets of the World, 1986–7* (Annapolis: Naval Institute Press, 1986), p. 119.
70. Britain is perhaps the most notable and long-standing case of conflict over maritime air power. See G. Till, *Air Power and the Royal Navy 1914–45* (London: Jane's, 1979); and Grove, *Vanguard to Trident*. In Italy the navy was only allowed to operate fixed wing aircraft by a change in the law in 1987. See *Jane's Defence Weekly*, 17 October 1987, p. 852.
71. M.A. Libbey, 'Airships in naval operations – a US evaluation', *Naval Forces* Vol. V. No. VI, 1984. N. Polmar, *The Ships and Aircraft of the US Fleet*, pp. 445–6 contains a useful summary of U.S. manned airship developments; for problems with the programme see *International Defense Review*, June 1987, p. 809.
72. E.D. Cooper and S.M. Shaker, 'Soaring into the next century, new directions for US Aerospace', *Journal of Defense and Diplomacy Study No. 4*, September 1988, p. 15. This study has also been drawn upon for later comments in this section. *Jane's Defence Weekly*, 19 November 1988, had a section on the new LRAACA's engines on p. 1289.
73. See *Jane's Defence Weekly*, 17 October 1987, pp. 852 for the latest Harrier developments.
74. Bill Sweetman, *Aircraft 2000: The Future of Aerospace Technology* (London: The Military Press, 1984), p. 83.
75. For consideration of more advanced designs see ibid., pp. 83–9, and 'Advanced naval STOVL aircraft under study', *International Defense Review*, January 1983.

Notes and references

76. B. Cooper, 'The V-22 Osprey tilt rotor, an aircraft for all services', *Navy International*, January 1988, pp. 22–5; *The V-22 Osprey* (Bell-Boeing brochures); for Osprey in the ASW role see also *Flight International*, 31 October 1987, p. 29. For suggestions for more advanced VTOL designs, see *Flight International*, 17 January 1987, p. 27. Sweetman, op. cit. illustrates a similar Vought VTOL concept on p. 90.
77. British Aerospace, *Sky Hook*, 1987, p. 1.
78. Ibid., pp. 14–15, 18–19.
79. See note 73.
80. F.D. Kennedy, 'RPVs at sea', *Naval Forces*, Vol. VI, No. V, 1985.
81. See Polmar, *Ships and Aircraft*, pp. 447–8 for discussion of unmanned LTA vehicles.
82. See Compton Hall, op. cit., pp. 98–100 for SPUR and discussion of other robotic projects. The Friedman quotation (and more on UUV) is in US Naval Institute *Proceedings*, August 1988, pp. 123–4.
83. Quoted in 'Satellites and naval operations', *Naval Forces*, Vol. VI, No. 5, 1985, p. 66, an important source for the rest of this section.
84. House Armed Services Committee testimony quoted in T.J. Richelson and Desmond Ball, *The Ties That Bind* (London: Allen and Unwin, 1985), p. 218; chapter 9, on 'Ocean Surveillance', is an excellent source on this sensitive subject.
85. N. Friedman in US Naval Institute *Proceedings*, May 1987, p. 93.
86. For a survey of Soviet ocean surveillance satellites see *Jane's Spaceflight Directory*, 1986, pp. 306–7.
87. In the Navy Review issue of US Naval Institute *Proceedings*, p. 78.
88. N. Friedman, *Modern Warship Design and Development*, (Greenwich: Conway, 1979), p. 45. This is still the finest survey of the subject and required reading for anyone who wishes to understand the dynamics of the design of modern surface warships.

Chapter 7 The international and legal context

1. *Discriminate Deterrence: Report of the Commission on Integrated Long Term Strategy*, Memorandum for the Secretary of Defense and the Assistant to the President for National Security Affairs, January 1988.
2. J. Dean, *Watershed in Europe: Dismantling the East-West Military Confrontation* (Lexington, Mass.: Lexington Books, 1987), pp. xiv–xv.
3. The author has assisted Windass in this work. See the two volumes 'Common Security in Europe', published by the Foundation for International Security in 1987. A book version entitled *The Crucible of Peace* was to be published in 1988 by Brasseys of London.
4. C. Hall, *Britain, America and Arms Control 1921–37* (London: Macmillan, 1987), p. 218.
5. L. Freedman, 'Arms control at sea', *Naval Forces*, Vol. VI, No. III, 1985.
6. J. Borawski, 'Risk reduction at sea: naval confidence building measures', *Naval Forces*, Vol. VIII, No. I, 1987. For the best brief summary of Stockholm see the same author's 'Accord at Stockholm', *Bulletin of Atomic Scientists*, December 1986, pp. 34–6.

Notes and references

7. Borwaski (1987), op. cit., p. 24.
8. For recent Soviet proposals see J. Borawski and E. Whitlow, 'A Nordic zone of peace?', *Naval Forces*, Vol. VIII, No. VI, 1987.
9. 'The Soviet–British incidents at sea agreement', *Naval Forces*, Vol. VIII, No. I, 1987; and 'Agreement between the USA and the USSR on the prevention of incidents on and over the high seas', pp. 166–7 of J. Goldblat *Arms Control Agreements: A Handbook* (London: Taylor and Francis, 1983). For the possibilities for East–West naval arms control measures see J.R. Hill, *Arms Control at Sea* (London: Routledge, 1988) and J.R. Hill and E.J. Grove, 'Naval co-operative security', in *Common Security in Europe* (Adderbury: Foundation for International Security, 1988).
10. K. Booth, *Law, Force and Diplomacy at Sea* (London: Allen and Unwin, 1985), p. 209.
11. Sir James Cable, 'Gunboat diplomacy's future', US Naval Institute, *Proceedings*, August 1986, p. 38.
12. Ibid., p. 40.
13. Booth, op. cit., pp. 181–2.
14. See succeeding volumes of the official Naval review issue of US Naval Institute *Proceedings*, published every May.
15. For surveys of Soviet operations see B. Dismukes and J.M. McConnell, *Soviet Naval Diplomacy* (New York: Pergamon, 1979); and B.W. Watson, *Red Navy at Sea: Soviet Naval Operations on the High Seas 1956–80*, (Boulder: Westview, 1982).
16. For an excellent examination of recent Soviet thinking on this see S. Shenfield, *The Nuclear Predicament: Explorations in Soviet Ideology*, (London: Royal Institute of International Affairs, 1987), especially p. 43.
17. Dismukes and McConnell, op. cit., p. 299.
18. E.N. Luttwak, *The Political Uses of Sea Power* (Baltimore: Johns Hopkins University Press, 1974), p. 52.
19. H. Coutau-Beggarie, 'The role of the French navy in French foreign policy', *Naval Forces*, Vol. VII, No. VI, 1986.
20. Cable, op. cit., p. 41.
21. For a stimulating critique of traditional British navalism see C. Coker, *A Nation in Retreat? Britain's Defence Commitment* (London: Brasseys, 1986), pp. 64–75. See also the final chapter of E.J. Grove, *Vanguard to Trident*.
22. C.W. Weinberger, 'NATO and the global maritime challenge', *SEALINK 86*, SACLANT, Norfolk, Virginia, 1986, p. 3.
23. Ibid., p. 4.
24. Ibid.
25. A.H. Cordesman, *The Iran–Iraq War and Western Security, 1984–87* (London: Jane's, 1987), pp. 132–4.
26. *Jane's Defence Weekly*, 22 August 1987, p. 296. Also see *The Observer*, 16 August 1987.
27. *The Guardian*, 12 August, 1987.
28. *Jane's Defence Weekly*, 19 September 1987, p. 581; 26 September 1987, pp. 263, 272–3; 24 October 1987, pp. 927, 941; *Daily Telegraph*, 22 January 1988; *The Times*, 25 October 1988; see also *The*

Notes and references

War In the Gulf: Background Brief (London: RUSI, June 1988) and also E.J. Grove, 'Towards a western European navy', *Naval Forces*, January 1988.
29. Colin Gray, 'National interests and the Pacific – implications for NATO', *Sealink 1986*, p. 36.
30. I.P.S.G. Cosby, 'Maritime Aspects of Japanese Security', *Naval Forces*, Vol. IV, No. V, 1983.
31. Quoted in B. Hahn, 'PRC policy in maritime Asia', *Journal of Defense and Diplomacy*, June 1986, p. 20.
32. F. Jiardano, 'The Chinese navy', *Naval Forces*, Vol. VIII, No. II, 1987, p. 198.
33. A.J. Tellis, 'India's naval expansion – structure, dimensions and context', *Naval Forces*, Vol. VIII, No. V, 1987, p. 45.
34. Ibid., p. 49.
35. *Review of Australia's Defence Capabilities*, Report to the Minister for Defence by Mr. Paul Dibb (Canberra: Australian Government Publishing Service, 1986), p. 1.
36. Ibid., pp. 3–4.
37. I owe the term to one of Australia's brightest young officers, Lt. Cdr. James Goldrick RAN, the noted naval historian.
38. Dibb Report, p. 8.
39. 'Viewpoint' by New Zealand's defence minister in *Jane's Defence Weekly*, 14 November 1987, p. 1119.
40. Cable, op. cit., p. 40.
41. Luttwak, op. cit., chap. 1.
42. See Booth, op. cit., chapter 1 for a useful survey of the background to UNCLOS III and the provisions of the Convention.
43. Ibid., p. 34.
44. Ibid., p. 35.
45. Barry Buzan, 'A sea of troubles', Adelphi Paper No. 143 (London: International Institute for Strategic Studies, Spring, 1978).
46. Ibid., pp. 4–15.
47. See the 'International Law' sections of *Britannica Book of the Year, 1985–87* for useful annual summaries of international legal disputes.
48. For a description of these 'Freedom of Navigation' operations see H. Parkes, 'Crossing the line', US Naval Institute *Proceedings*, November 1986, pp. 43–5.
49. N.M. Hunnings, 'Law', *Britannica Book of the Year*, 1986, p. 291.
50. Booth, op. cit., p. 37.
51. Ibid., p. 38.
52. Ibid., pp. 44–5. Booth introduced the notion of the psycho-legal boundary in his article 'Naval strategy and the spread of psycho-legal boundaries at sea', *International Journal*, Summer, 1983.
53. L.M. Alexander, 'The ocean enclosure movement: inventory and prospect', *San Diego Law Review*, Vol. 20, No. 3, pp. 561–94.
54. Booth (1985), op. cit., p. 168.
55. For a discussion of some of the perceived obstacles to setting up a UN

Notes and references

Gulf peacekeeping force see J. McCoy, 'Naval mirage in the Gulf', *The Times*, 5 July 1988.
56. Cable, op. cit., p. 38.
57. Booth (1985), op. cit., 170.

Chapter 8 Economic stimuli and constraints

1. Figures calculated from relevant volumes of *Britannica Book of the Year*' also see R. Hill, *Maritime Strategy for Medium Powers* (London: Croom Helm, 1986), p. 31.
2. See various statements by the British Maritime League. For an excellent in depth analysis of the decline of the British merchant fleet see *Why the Ships Went*, (London: British Maritime Charitable Foundation, 1986). For a well-written critique of British attitudes see Sir James Cable, 'Closing the British seas', in *Marine Policy*, Vol. 11, No. 2, April 1987.
3. Statement made in 1982. Quoted in 'Resource wars, the myth of American vulnerability', *Defence Monitor*, Vol. XIV, No. 9, 1985.
4. The US Navy League has produced such figures; see for example *The Almanac of Sea Power 1983*, p. 196.
5. See note 3.
6. *Times Atlas of the Oceans*, (London: Times Books, 1983), p. 144.
7. Hill, op. cit., p. 42.
8. Ibid., p. 42.
9. Ibid., pp. 43–4.
10. H.W. Maull, *Raw Materials, Energy and Western Security* (London: Macmillan, 1984), p. 110.
11. 'Japan's energy security to 2000 – the dawn of the age of multiple energies', in R. Belgrave, C.K. Ebinger and H. Okino, *Energy Security to 2000* (London: Gower; Boulder, Colorado: Westview), 1987, p. 57.
12. E. Solem and A.F.G. Scanlon, 'Oil and natural gas as factors in strategic policy and action', in A.H. Westing (ed.) *Global Resources and International Conflict* (SIPRI and Oxford University Press, 1986).
13. *Times Atlas of the Oceans*, p. 189.
14. S.R. Peterson and J.M. Teal, 'Ocean Fisheries', in *Global Resources and International Conflict*, supra., note 11.
15. 'Shipbuilding', *Britannica Book of the Year* 1986, pp. 256–7 gives the world ranking.
16. 'The maritime role in strategic deterrence and crisis management', *Sealink 86*, (Norfolk, Virginia: SACLANT, 1986).
17. Cable, op. cit., p. 94.
18. *Sealink 86*, op. cit.
19. Maull, op. cit., p. 9.
20. *The War in the Gulf: Background Brief* (London: RUSI, June 1988), pp. 20–1.
21. Hill, op. cit., p. 47.
22. S.C. Truver, 'Gramm-Rudman and the future of the 600-ship fleet', US Naval Institute *Proceedings*, May 1987, p. 123.

Notes and references

23. *Washington Post*, 8 August 1986, p. A14, quoted Truver, op. cit., p. 120.
24. Ibid., p. 123.
25. *Jane's Defence Weekly*, 20 February 1988, p. 285.
26. Philip Pugh, *The Cost of Seapower: The Influence of Money on Naval Affairs from 1815 to the Present Day* (London: Conway, 1986). A good summary of his major points appeared in *Naval Forces*, No. III, 1987.
27. Pugh, op. cit., p. 314.
28. Ibid., p. 363.
29. Ibid., pp. 386–7.
30. See E.J. Grove, *Vanguard to Trident, British Naval Policy since World War Two* (Annapolis: Naval Institute Press; London: Bodley Head, 1987), for repeated demonstration of this basic rule with regard to the post 1945 Royal Navy.
31. Pugh, op. cit., pp. 387–8.
32. Ibid., p. 388.

Chapter 9 Navies in peace

1. R. Hill, 'Control of the exclusive economic zone', *Naval Forces*, No. VI, 1985. See also the same author's *Maritime Strategy for Medium Powers*, chap. 6.
2. J.D. Harbron, 'Who shall defend the Canadian Arctic?', *Jane's Naval Review*, 5th Year of Issue, (London: Jane's, 19??), pp. 151–7.
3. See E.J. Grove, *Vanguard to Trident, British Naval Policy since World War Two* (Annapolis: Naval Institute Press; London: Bodley Head, 1987), especially chaps 6 and 10.
4. R. Villar, *Piracy Today: Robbery and Violence at Sea Since 1980* (London: Conway, 1985), pp. 10–11.
5. Ibid.
6. 'Navies in a terrorist world', *Jane's Naval Review*, 5th Year of Issue, p. 172.
7. For descriptions of these operations see US Naval Institute *Proceedings*, especially for annual reviews of naval operations published in May of every year. For the major Libyan raid see also *Aviation Week and Space Technology*, 31 March 1986.
8. Ibid.
9. J.D. Watkins, 'The Maritime Strategy' in *The Maritime Strategy* supplement to US Naval Institute *Proceedings*, January 1986, p. 8.
10. S.S. Roberts, 'The October 1973 Arab–Israeli War', in B. Dismukes and J. McConnell, *Soviet Naval Diplomacy* (New York and London: Pergamon, 1979).
11. For a survey of the Soviet Navy's operations see Dismukes and McConnell, op. cit.; B. Watson and S.M. Watson, *The Soviet Navy Strengths and Liabilities* and B.W. Watson, *Red Navy at Sea: Soviet Naval Operations on the High Seas*, (Boulder, Colo.: Westview, 1982 and 1986, respectively).
12. K. Booth, *Navies and Foreign Policy* (London: Croom Helm, 1977); and Sir James Cable, *Gunboat Diplomacy 1919–79* (London: Macmillan, 1981).
13. Booth, op. cit., pp. 18–20.

Notes and references

14. Sir James Cable, 'Showing the flag: past and present', *Naval Forces*, No. III, 1987, p. 38. This important article is effectively a supplement to the book, *Gunboat Diplomacy*.
15. Cable (1981), op. cit., p. 38.
16. Cable (1987), op. cit., p. 38.
17. Cable (1981), op. cit., p. 57.
18. Ibid., p. 67.
19. Ibid., p. 81.
20. Ibid., p. 117.
21. For more discussion of reach see J.R. Hill, *Maritime Strategy for Medium Powers*, chap 10.
22. Watkins, op. cit., p. 8.
23. Vice-Admiral Sir Geoffrey Dalton, 'NATO's maritime strategy', in *NATO and US Maritime Strategy: Diverging Interests or Co-operative Effort?* (Oslo: Norwegian Atlantic Committee, 1987), p. 42.
24. Ibid., pp. 42-3.
25. Ibid., p. 43.
26. Vice-Admiral H.C. Mustin, 'Maritime strategy from the deckplates', US Naval Institute *Proceedings*, September 1986, p. 35.
27. For further development of these ideas see E.J. Grove, 'The maritime strategy and crisis stability', *Naval Forces*, Vol. VIII, No. VI, 1987.

Chapter 10 Navies in war

1. For French nuclear doctrine and the role of submarines within it see D.S. Yost, *France's Deterrent Posture Part 1 – Capabilities and Doctrine* (London: IISS Adelphi Paper, 1984). A full description of Britain's targeting all its warheads on Moscow was broadcast in J. Barry, 'Our bomb, the secret story', Channel 4 Television, April 1986.
2. For an excellent survey of the history of NATO's sealift requirement see J.J. Sokolsky, 'Seapower in the nuclear age', unpublished doctoral dissertation presented to Harvard University, August 1984. Figures for current sealift requirement from unclassified EASTLANT briefing given at Northwood, 30 January 1987.
3. Northwood briefing, ibid.
4. The 1,000 shipping movements in the first month were discussed by Assistant Under-Secretary (Programmes), R.C. Mottram, in testimony to the House of Commons Defence Committee, 3 February 1988, House of Commons Session 1987-88, Defence Committee, Sixth Report, *The Future Role and Size of the Royal Navy Surface Fleet*, p. 8.
5. *Sealink 86*, (Norfolk, Va.: SACLANT, 1986), p. 47. Cargo figures from Vice-Admiral Sir Geoffrey Dalton, 'NATO's maritime strategy', in E. Ellingsen (ed.) *NATO and US Maritime Strategy: Diverging Interests or Co-operative Effort* (Oslo: Norwegian Atlantic Committee, 1987), p. 44.
6. *Sealink 86*, op. cit.; Dalton, op. cit.
7. C.H. Whitehurst, *The U.S. Merchant Marine* (Annapolis: Naval Institute Press, 1983), p. 131; Dalton, op. cit., pp. 44-6.

Notes and references

8. *Sealink 86*, op. cit., p. 47.
9. The organization of the Joint Chiefs of Staff, *United States Military Posture, Fiscal Year 1987*, p. 68.
10. 'Analysis of the September 1987 sealift ship list', unclassified section of 1987 PBOS Report.
11. Ibid. For a longer discussion of sealift issues see E.J. Grove's paper 'The merchant fleet and deterrence', prepared for the British Maritime Charitable Foundation, 1988.
12. See notes 11 and 31, chapter 1.
13. See E.J. Grove, *Vanguard to Trident, British Naval Policy since World War Two* (Annapolis: Naval Institute Press; London: Bodley Head, 1987); N. Friedman, *The Post-War Naval Revolution*; and G. Till and E. Grove, 'Anglo American naval strategy 1945–67 – the impact of the Cold War, NATO and nuclear weapons', in *Maritime Strategies of the Great Powers: Anglo-American Perceptions* (London: Macmillan, 1989).
14. 'Finally seizing the initiative opens the way to apply direct pressure on the Soviets to end the war on our terms – the new goal of our strategy once deterrence has failed.' Admiral J.D. Watkins, 'The Maritime Strategy', in *The Maritime Strategy*, supplement to US Naval Institute *Proceedings*, January 1986, p. 11.
15. B.R. Posen, 'Inadvertent nuclear war? Escalation and NATO's northern flank', *International Security*, Vol. 7, No. 2, Fall 1982, reprinted in S.E. Miller (ed.) *Strategy and Nuclear Deterrence* (Princeton, NJ: Princeton University Press, 1984).
16. P.H. Nitze and L. Sullivan Jr., *Securing the Seas; The Soviet Naval Challenge and Western Alliance Options* (Boulder Colo.: Westview, 1979), p. 373.
17. J.R. Hill, *Anti-Submarine Warfare* (Shepperton: Ian Allan, 1984), pp. 98–103.
18. Nitze and Sullivan, op. cit., pp. 370–1.
19. The scenario of the 'Ocean Safari' Exercise in 1987.
20. 'A survey of western ASW in 1985', *International Defense Review*, No. 10, 1985, p. 1597.
21. Translated from Vice-Admiral Sir Geoffrey Dalton, *Atlanterhavet: NATO's Strategiske Planlegging* (Norwegian Atlantic Committee, 1986), p. 7.
22. Dalton, op. cit., p. 40.
23. Ibid., p. 42.
24. See for example the comprehensive report on 'Ocean Safari 83' by Martin Horseman in *Armed Forces*, October 1983. The author observed Exercise 'Teamwork 88' in person.
25. Dalton, op. cit., p. 43.
26. Major-General D.J. Murphy, 'Role of the U.S. marine corps', and Vice-Admiral Torolf Rein, 'Reinforcing the northern flank. The case of Norway', in E. Ellingsen (ed.), *Reinforcing the Northern Flank* (Oslo: Norwegian Atlantic Committee, 1987) pp. 57 and 34 respectively.
27. *Military Posture*, FY1987, p. 69.
28. Murphy, op. cit., p. 54.

Notes and references

29. Brigadier R.J. Ross, 'Organisation of UK/NL landing force', in Ellingsen, *Reinforcing the Northern Flank*, op. cit., pp. 66–7.
30. Vice-Admiral Torolf Rein, 'The situation in the Norwegian Sea and the Norwegian naval interests', Ellingsen, *NATO and U.S. Maritime Strategy*, op. cit., p. 57.
31. Dalton, op. cit., pp. 42–3. For recent discussion of Mediterranean strategy see the proceedings of the 1987 Conference of the International Institute for Strategic Studies held at Barcelona, *Prospects for Security in the Mediterranean* (Adelphi Papers 229–231).
32. H.K. Ullman, 'The Pacific and U.S. naval policy, *Naval Forces*, No. VI, 1985, p. 42. For the importance of Pacific thinking in the evolution of the Maritime Strategy see J.B. Hattendorf, 'The evolution of the Maritime Strategy', *Naval War College Review*, Summer 1988.
33. See P.X. Kelley, 'The amphibious warfare strategy', *Maritime Strategy* brochure, p. 26.
34. This analysis is based on T.L. McNaugher's excellent study *Arms and Oil: U.S. Military Strategy and the Persian Gulf* (Washington: Brookings, 1985, pp. 55–61).
35. Military Posture, FY1987, p. 69. There are thirteen modern MPS ships currently on charter, five in the Indian Ocean, four in the Pacific and four in the Atlantic. See *Combat Fleets of the World*, 1988, pp. 803–4.
36. Ibid., and McNaugher, op. cit., p. 67–8.
37. Ibid.; see also pp. 78–9.
38. Ibid.; chapter 2 gives a good assessment of the Soviet threat.
39. Professor L. Martin's analysis in *The Sea in Modern Strategy* (London: Chatto and Windus, 1967) was heavily conditioned by this perspective.
40. A.H. Cordesman, *The Iran–Iraq War and Western Security* (London: Jane's, 1987), pp. 46–7.
41. Ibid., p. 68.
42. Ibid., especially pp. 46–56.
43. John Wise, 'Maritime electronic warfare and its applicability to the protection of fleet support units or merchant vessels operating in hostile waters', *Proceedings of ATI European Naval Forecast Conference, London*, 14–15 May 1987.
44. Cordesman, op. cit., p. 116 quoting US Congressional testimony.
45. Ibid., pp. 132–4.
46. A.H. Cordesman, 'Western seapower enters the Gulf', *Naval Forces*, II, III and IV, 1988 gives a comprehensive and most useful account from the American perspective of the growing outside naval involvement. He understates, however, the importance of European 'concertation'; see chapter 7.
47. N. Friedman in US Naval Institute *Proceedings*, June 1988, p. 119.

Chapter 11 The future of seapower – three theoretical frameworks

1. For Corbett's and other contemporary views on Mahan see E.J. Grove's introduction to Corbett's *Some Principles of Maritime Strategy*

Notes and references

(Annapolis: US Naval Institute; London: Brasseys, 1988).
2. For the scope of Mahan's writings see John B. and Lynn C. Hattendorf, *A Bibliography of the Works of Alfred Thayer Mahan* (Newport, Rhode Island: Naval War College Press, 1986). The Hattendorfs list over fifty English editions and printings of *The Influence of Sea Power Upon History 1660–1783*, (first published by Little Brown and Co., Boston and Sampson Low Marston, London in 1890) and fourteen translations.
3. Ibid., p. 25.
4. Ibid., p. 26.
5. See the House of Commons Defence Committee's Report, *The Defence Requirement for Merchant Shipping and Aircraft*, June 1988, and the author's report for the British Maritime Charitable Foundation, *The Merchant Fleet and Deterrence*.
6. A.T. Mahan, *The Influence of Sea Power upon History 1660–1783* (Boston: Little Brown, 1890), p. 23.
7. Ibid.
8. See for example Graham Searjeant, 'The rising tide of protectionism', *Britannica Book of the Year 1988*, pp. 172–3.
9. J.R. Hill, *Maritime Strategy for Medium Powers* (London: Croom Helm, 1986), chap. 10.
10. Mahan, op. cit., p. 26.
11. Ibid.
12. Ibid., pp. 28–9.
13. Ibid., p. 29.
14. Lord Fraser at a meeting of the Chiefs of Staff, COS(50) 46th Meeting, DEFE4/29.
15. Mahan, op. cit., p. 29.
16. Ibid., p. 40.
17. Ibid., p. 43.
18. Ibid., p. 45.
19. Ibid., p. 53.
20. Ibid., p. 67.
21. Philip Pugh, *The Cost of Sea Power: The Influence of Money on Naval Affairs from 1815 to the Present Day* (London: Conway, 1986), p. 23.
22. Mahan, op. cit., p. 82.
23. *Jane's Fighting Ships, 1987–88*, respectively, pp. 494, 240, and 49.
24. Hill, op. cit., pp. 42–3.
25. Mahan, op. cit., p. 88.
26. K. Booth, *Navies and Foreign Policy* (London: Croom Helm, 1977), especially p. 16.
27. Ibid., pp. 15–16.
28. Ibid.
29. Ibid., p. 16.
30. Sir James Cable, *Gunboat Diplomacy 1919–1979* (London: Macmillan, 1981); see chap 9 in this volume.
31. Commodore Richard Cobbold, Director, Defence Concepts to the Royal United Services Institute, 18 November 1980.
32. I owe the 'three circles concept' to my colleague in the Foundation for

Notes and references

International Security, Stan Windass.
33. See, for example, K. Booth, *Navies and Foreign Policy* (London: Croom Helm, 1977); Hill, op. cit.; M.W. Janis, *Sea Power and the Law of the Sea*, (Lexington, Mass: Heath, 1976); and M.A. Morris, *Expansion of Third World Navies* (London: Macmillan, 1987). Two important recent articles on this subject are Nien-Tsu Alfred Hu and J.K. Oliver, 'A framework for small navy theory' *Naval War College Review*, Spring 1988 and S.W. Haines, 'Third World navies: myths and realities', *Naval Forces*, April 1988.
34. Haines, op. cit., p. 24.
35. Morris, op. cit., p. 25.
36. Ibid.
37. Ibid.
38. Ibid., p. 33.

Acronyms and abbreviations

AAW	anti-air warfare
ABM	anti-ballistic missile
ACV	air cushion vehicle
ADA	Action Data Automation System
ADAWS	Action Data Automation and Weapons System
ADC	acoustic device countermeasures
AEW	airborne early warning
AGI	auxiliary ships
AIO	Action Information Organization
AMRAAM	advanced medium-range air-to-air missile
ANS	*Anti Navires Supersonique*
ANZDCC	Australian/New Zealand Defence Consultative Committee
ARMS	Advanced Remote Minehunting System
ASATS	anti-satellite weapons
ASROC	anti-submarine rocket
ASUW	anti-surface ship warfare
ASW	anti-submarine warfare
ASW-SOW	Anti-Submarine Warfare Stand-Off Weapon
ATSA	advanced tactical support aircraft
BASH	barge aboard ship
Bo-Ro	general purpose Ro-Ro bulk carrier container ship
CAAIS	Computer-Assisted Action Information System
CACS	Computer-Assisted Command System
CBM	confidence building measures
CCTS	Closed-Cycle Thermal System
CDE	Conference on Confidence and Security Building Measures and Disarmament in Europe
CENTCOM	Central Command
CEP	circular error profitability
C^3I	command, control, communications and intelligence

Acronyms and abbreviations

CIC	Combat Information Centre
CINCHAN	Commander-in-Chief Channel
CINCSOUTH	Commander-in-Chief South
CIWS	Close-In Weapons System
CLGP	cannon-launched guided projectile
CMEA	Council for Mutual Economic Assistance
CODAD	combined diesel and diesel turbine
CODAG	combined diesel and gas turbine
CODLAG	combined diesel electric and gas turbine
CODOG	combined diesel or gas turbine
COGAG	combined gas and gas turbine
COGOG	combined gas or gas turbine
CONMAROPS	Concept of Maritime Operations
CSBM	confidence and security building measures
CTOL	conventional take-off and landing
CVBG	carrier battle groups
CVV	mid-size carriers
DASH	Drone Anti-Submarine Helicopter
DE	destroyer escort
DIFAR	directional low-frequency analysis and recording
DSMAC	digital scene matching area correlator
EASTLANT	Eastern Atlantic Command
ECM	electronic countermeasures
EEZ	Exclusive Economic Zone
ELF	extremely low frequency
Eorsat	Electronic Ocean Surveillance Satellite
FFG	guided missile frigate
HELRAS	helicopter long-range sonar
ICBM	intercontinental ballistic missile
INF	intermediate nuclear forces
ISAR	inverse synthetic aperture radar
ISO	International Standardization Organization (based in Geneva)
LASH	lighter aboard ship
LCAC	landing craft air cushion
LCAT	low-cost acoustic torpedo
LHA	landing helicopter assault
LHD	landing-ship helicopter dock
LNG	liquid natural gas
LOFAR	low-frequency analysis and recording
LPD	landing platform dock
LPG	liquid petroleum gas
LPH	landing platform helicopter
LRAACA	long-range air ASW-capable aircraft

Acronyms and abbreviations

LSD	landing ship docks
LTA	ligher than air
MAD	magnetic anomaly detection
MCM	mine countermeasures
MEF	Marine Expeditionary Force
MESAR	multi-function electronically scanned array radar
MIRV	multiple independently targeted re-entry vehicle
MRNA	Multi-Role Naval Auxiliary
NATO	North Atlantic Treaty Organization
NAVSTAR	navigation system using tuning and ranging
NTDS	naval tactical data system
OAU	Organization for African Unity
OECD	Organization for Economic Co-operation and Development
PBV	post-boost vehicle
PLARB	Russian equivalent of SSBN
PREPO	pre-positioning
ROE	rules of engagement
Ro-Ro	roll-on roll-off
Rorsat	Radar Ocean Surveillance Satellite
RPV	remotely piloted vehicle
SACEUR	Supreme Allied Commander Europe
SACLANT	Supreme Allied Commander Atlantic
SALT	Strategic Arms Limitation Talks
SAM	surface-to-air missile
SAR	synthetic aperture radar
SCEPS	stored chemical energy propulsion system
SCS	sea control ships
SDI	strategic defense initiative
SEALS	sea-air-land special forces
SES	surface effect ship
SHF	super high frequencies
SIOP	Strategic Integrated Operational Plan
SLAT	surface-launched air-targeted
SLBM	submarine-launched ballistic missile
SLCM	sea-launched cruise missile
SLCSAT	Submarine Laser Communication Satellite System
SMCS	Submarine Command System
SOSUS	Sound Surveillance System
SPUR	Seicon Patrolling Underwater Robot
SSBN	nuclear-powered ballistic submarine
SSM	surface-to-surface missile
SSN	nuclear-powered submarine
START	Strategic Arms Reduction Talks

Acronyms and abbreviations

STOVL	short take-off/vertical landing
SubACS	Submarine Advanced Combat System
SWATH	small waterplane twin-hull
SWCM	special warfare craft medium
T-AGOS	ocean surveillance ship
TERCOM	terrain contour-matching
teu	twenty-foot equivalent units
TLAM	Tomohawk cruise missile
TOAD	towed offboard active decoy
UEP	underwater electrical potential
UHF	utlra high frequencies
ULCC	ultra large crude carrier
UNCLOS III	Third United Nations Conference on the Law of the Sea
UUV	unmanned underwater vehicle
VLAD	vertical land array DIFAR
VLCC	very large crude carrier
VLF	very low frequencies
VSS	V/STOL support ships
V/STOL	vertical/short take-off and landing
VTOL	vertical take-off and landing
WEU	Western European Union
WIG	'wing in ground'
WTO	Warsaw Treaty Organization

Index

acoustic device countermeasures (ADC) 75
Action Data Automation System (ADA) 90
Action Data Automation Weapons System (ADAWS) 90–1
Action Information Organization (AIO) 90–1
Adcock, C.C.F. 80
advanced medium-range air-to-air missile (AMRAAM) 86, 141
Advanced Remote Minehunting System (ARMS) 72
advanced tactical support aircraft (ATSA) 141
Aegis system 83
air cushion vehicle (ACV) 118
airborne early warning (AEW) 86, 141, 144
aircraft 138–45; advantages 139; and airborne early warning facilities 144; and anti-submarine warfare 140–1; drawbacks 139–40; and maritime strategy 14; short take-off/vertical landing 141–3; and skyhook 143; threat of 81; vertical take-off and landing 142–4
aircraft carriers 124–8; maritime strategy 14; mid-size (CVV) 127
Albania 240
Alexander, L.M. 170
Algeria 175, 239
Amery, L. 5
amphibious force 211
amphibious vessels 129–30
amphibious warfare 49; maritime strategy 22–3; and military power 48–51
Angola 240
Anti Navires Supersonique (ANS) 66
anti-satellite weapons (ASATS) 148
anti-air warfare (AAW) 61, 81–7; future of 235; and surface ships 100–28 *passim*; weapons for 139
anti-ballistic missile (ABM) 87, 148, 199
anti-submarine rocket (ASROC) 76, 107, 113
anti-submarine warfare (ASW) 15, 72–81; future of 235; and nuclear warfare 200–1; and surface ships 99–132 *passim*; techniques of 53; weapons for 63, 65, 139
Anti-Submarine Warfare Stand-Off Weapon (ASW-SOW) 76
anti-surface ship warfare (ASUW) 61; and surface ships 102–26 *passim*; techniques of 74–5; weapons for 61–4, 139
Argentina 176, 239, 240
Australia 165–6, 175–6, 238
Australian/New Zealand Defence Consultative Committee (ANZDCC) 166
auxiliary ships (AGI) 55

Baggett, Admiral L. 203
Bahrain 240
Bangladesh 239

Index

barge aboard ship (BASH) 37
battlefleets 25–6
battleships 103–4
Belgium 175–6, 238
Berlin Crisis 57
blockade 15, 16
Booth, K.: and international conflict 156–7, 171; and order at sea 167, 170; peacetime uses of navies 194; and UNCLOS III 169–70; and use of the sea 232–5
Borawski, J. 155
Brazil 175–6, 178, 239
Britain: confrontation with Union of Soviet Socialist Republics 156; economic use of the sea 172, 175–6; foundation of sea power 224, 229; as global force 238; merchant shipping 4, 40–1; and mine countermeasures 99; navies in peacetime 188; shipbuilding 178; surface vessels 105–6, 110–11, 126, 130–1; torpedoes 65, 73–4; World War II strategy 50
Brunei 240
Bulgaria 239
Burma 240
Butler, E. 118–19
Buzan, B. 168

Cable, Sir J.: and international conflict 156–7, 166, 171; merchant shipping 179; peacetime uses of navies 194–5; and smaller navies 160; and use of the sea 233–4
Cameroons 240
Canada 107, 136–7, 175–6, 188–9, 202–3, 240
cannon-launched guided projectile (CLGP) 62
Cape Matapan 14
carrier battle group (CVBG) 207–8, 213
Central Command (CENTCOM) 214
Chernavin, Admiral S. 193

Chile 104, 176, 239
China 40–1, 43, 108–9, 131, 163–4, 178, 238
circular error profitability (CEP) 95
Clausewitz, K. von 18
Close-In Weapons System (CIWS) 82, 106, 109
Closed-Cycle Thermal System (CCTS) in torpedoes 74–5
co-operation: in Persian Gulf 162; in security 154–5
Coast Guards 187–8
coastal bombardment 87–8
coastlines and foundations of sea power 225
Colombia 239
Combat Information Centre (CIC) 90, 115
Combined diesel and diesel turbine (CODAD) 123
combined diesel electric and gas turbine (CODLAG) 106, 123
combined diesel and gas turbine (CODAG) 121
combined diesel or gas turbine (CODOG) 121
combined gas and gas turbine (COGAG) 123
combined gas or gas turbine (COGOG) 123
command and control of weapons systems 89–92
command of the sea 12–13, 19
Commander-in-Chief Channel (CINCHAN) 211
Commander-in-Chief South (CINCSOUTH) 212
communications: satellites 146, 148–9; submarines 133–4
Computer-Assisted Action Information System (CAAIS) 90
Computer-Assisted Command System (CACS) 90–1
Concept of Maritime Operations (CONMAROPS) 15, 209–13
confidence and security building measures (CSBM) 156, 197
constabulary navies 240

Index

containers *see* utilization
conventional take-off and landing (CTOL) 141–2
convoy 11–12; in future wars 209–10; operational laws of 19–21; and warfare 208–9; in World War I 18; in World War II 17–18, 19
Corbett, Sir J. 6, 10, 19, 221; and amphibious warfare 49; and constitution of fleets 24; and convoy 18; principle of maritime strategy 11–15, 23, 58; and uses of the sea 235
corvettes: definition 102; role of 112–13
Council for Mutual Economic Assistance (CMEA) 9, 174, 178
Coupar, A. 174
crisis management 196–8
cruisers: definition 102; role of 102–3
Cuba 239
Cunningham, Admiral A.B. 14
Cyprus 40

Dalton, Sir G. 203
Dean, J. 154
Denmark 175–6, 239
depth charges 77–8; nuclear 93
destroyers: definition 102; destroyer escort (DE) 113; role of 104–9
deterrence 187, 196
Dibb, P. 165–6
digital scene matching area correlator (DSMAC) 88
diplomatic role of navies 233–4
directed energy weapons 86–7
directional low-frequency analysis and recording (DIFAR) 80
Dominican Republic 240
Drone Anti-Submarine Helicopter (DASH) 145

Eames, M. 117
East Germany 239
Eastern Atlantic Command (EASTLANT) 203

Eberle, Admiral Sir J. 179
economic shipping during wartime 203–4
economic use of the sea 31–45; coal transport 34; dry bulk cargoes 34–5, 36–40. fishing and whaling 41–4; and flags of convenience 40–1; merchant shipping 41; passenger traffic 39; petroleum transport 31–3, 40, 44–5; roll-on roll-off 36–8; sea routes 38; trade 31–8; unit cargo vessels 37–8; unitization of cargo 35–7, 40; world trading fleet 39–40
economy: dependence on the sea 174–7; financial flows 179; strength of and sea power 226–8; of superpowers 153–4; and trade by sea 172–3; and use of the sea 172–83
Ecuador 239
Egypt 176, 239
electronic countermeasures (ECM) 216
Electronic Ocean Surveillance Satellite (Eorsat) 147
Esher, Lord 6
Ethiopia 240
Exclusive Economic Zone (EEZ) 42, 168, 170, 187

Falklands Islands 55; aircraft 85–6; missiles 68–9, 84; torpedoes in 63
fast-attack craft 99–100
ferries *see* passenger traffic
Finland 176, 240
Fisher, Admiral J.A. 26
fishing and dependence on the sea 178
fishing and whaling 41–4
flags of convenience 40–1
fleet support vessels 130–1
forward operations in warfare 210–11
foundations of sea power 221–32; coastlines 225; geographical position 224–5; and government

228–9; keys to 223–4; and merchant shipping 221–3; and national character 226–8; population 226; ranking of factors 229–32; territorial extent 225
France: economic use of the sea 175–6; as global force 238; maritime intervention 159; merchant shipping 4; submarines 135; surface vessels 105–6, 112, 125; torpedoes 75
Friedman, N. 149, 209
frigates: definition 102; role of 109–11

Gabon 175, 240
Gallipoli Campaign 48–9
Galtieri, L. 195
gas *see* petroleum
geographical position: and foundations of sea power 224–5; and military power 47–8
Ghana 240
global conflict: increase in 156; intervention 157–9
global force navies 237–8
Gorbachev, M. 154–5
Gorshkov, Admiral S. 12, 22, 24, 47, 87, 158
government and foundations of sea power 228–9
Gramm-Rudman-Hollings Act (1985) 180
Gray, C. 162
Greece 3, 40–1, 175–6, 239
Grenada, US Marines landings 157, 193
'guerre de course' 15–16, 21
Gulf, Arabian 68, 70, 160
gunboat diplomacy 192–6; kinds of 195; levels of 195–6
guns: anti-aircraft 81–3; naval 61, 62–3

Haines, S.W. 237
Hayward, Admiral T. 213
helicopter carriers 128
helicopter long-range sonar

(HELRAS) 80
Hill, Admiral J.R. 98, 174, 176, 180, 187, 223, 225, 230
Hong Kong 40–1

Iceland 240
India 108–9, 131, 164–5, 175–6, 238, 240
Indonesia 239
integrated weapons systems 89–91
intelligence-gathering satellites 147–8
intercontinental ballistic missile (ICBM) 94
intermediate nuclear forces (INF) 93
internationalization of sea 181–3, 233
intervention: by post-imperial powers 159–61; by superpowers 157–9, 161
inverse synthetic aperture radar (ISAR) 67
Iran 175–6, 239
Iran–Iraq war 160–1, 216–17; submarines 133
Iraq 239
Ireland 240
Israel 176, 239
Italy 75, 107, 175–6, 178, 238

Japan: amphibious warfare 49; economic use of the sea 175–6; fishing 42; foundation of sea power 227; merchant shipping 3, 40–1; as naval power 163; as regional force 238, 240; shipbuilding 178; surface vessels 106–7, 113

Kahn, H. 201
Kennedy, P. 4, 15
Kenya 240
Kuwait 160, 175, 240

Lambert, A. 23–4
landing craft air cushion (LCAC) 128
landing helicopter assault (LHA) 128

Index

landing platform dock (LPD) 128–9
landing platform helicopter (LPH) 54, 128
landing ship dock (LSD) 128–9
law of the sea 167–9, 170
Lebanon 175
legal disputes 168–9
Lehman, J. 124, 140
length-beam ratio 115–16
Leyte Gulf, battle of 26
Liberia 3, 40, 175
Libya 175, 239
lighter aboard ship (LASH) 37
Littwak, E. 159, 166
long-range air ASW-capable aircraft (LRAACA) 140
low-cost acoustic torpedo (LCAT) 64
low-frequency analysis and recording (LOFAR) 80
Lygachev, Y. (Russian politican) 156

MccGwire, M. 89
Mackinder, H. 4–5, 8
magnetic anomaly detection (MAD) 53
Mahan, A.T. 10, 15, 221; and convoy 16; definition of sea power 3, 4, 6; and economic uses of the sea 172; foundations of sea power 222–6, 228–9, 231–2; maritime strategy 22; and military power 46–7; and naval battles 13, 205
Malaysia 239
Marine Expeditionary Force (MEF) 214
maritime strategy 11–30; and aircraft 14; aircraft carriers 14; amphibious warfare 22–3; battlefields 25–6; blockade 15, 16; and command of the sea 12–13, 19; constitution of the fleet 24–6; and convoy 11–12, 16–19; definition of 11; 'fleet in being' 21–2; 'guerre de course' 15–16, 21; and naval battles 13–14; and nuclear weapons 14; operational laws 19–21
Martin, L. 57
Maull, H.W. 177
merchant shipping and foundations of sea power 221–3; *see also* economic use of the sea
Metcalf, Vice-Admiral J. 114
Mexico 176, 240
Middle East: and petroleum exports 31–2; in superpower war 213–15
Midway, battle of 13
military power 46–60; advantages of sea 54–5; amphibious warfare 48–51; efficiency of transport 47; geographical position 47–8; intelligence 55; and role of navies 233–4; and rules of engagement 56–7; and submarines 53–4, 55; in World War II 49–50
mines 61, 70–2; anti-submarine warfare 76–7; in Persian Gulf 161
mine countermeasures (MCM) 71–2, 99, 118, 128, 146, 161–2, 237
minesweepers 99
missiles 61, 65–70; anti-aircraft 81–3, 84–5
Morocco 239
Morris, M.A. 237–8
Multi-Role Naval Auxiliary (MRNA) 130
multi-function electronically scanned array radar (MESAR) 83–4
multihull craft 98, 117
multiple independently targeted re-entry vehicle (MIRV) 200
Mustan, Admiral H.C. 197

naval battles and maritime strategy 13–14
naval diplomacy 193–5
naval exercises 155–6
Naval Tactical Data System (NTDS) 90
navies: in peace 187–98; role of 232–6; typology for 236–40; and

use of the sea 232–3; in war 199–218
navigation system using tuning and ranging (NAVSTAR) 95
Netherlands, The 107, 175–6, 238
New Zealand 165, 176, 239
Nigeria 175–6, 239
North Atlantic Treaty Organization (NATO) 10, 15, 57, 228; areas of interest 160, 162; Concept of Maritime Operations 196–7, 209–10; convoys 208–9; economic shipping in wartime 203–4; and maritime strategy 205–8, 212, 235; and Rapid Reinforcement Plan 202
North Korea 239
North Yemen 240
Norway 41, 175–6, 239, 240
nuclear-powered ballistic submarine (SSBN) 89, 136, 154, 189–90, 201, 237–8
nuclear-powered submarine (SSN) 135–6, 238
nuclear warfare 199–202, 215
nuclear weapons 92–5; ballistic missiles 94–5; and maritime strategy 14; *pré-stratégique* strike 94; strikes against shore 93

ocean surveillance ship (T-AGOS) 78, 120
oil *see* petroleum
Oman 160, 240
Organization for African Unity (OAU) 157
Organization for Economic Co-operation and Development (OECD) 9, 174, 213

Pakistan 239, 240
Panama 3, 40, 175
passenger traffic 39
patrol vessels 98–9
peacetime uses of navies 187–98; and choice of vessels 189–90; coastguards 187–8; constabulary duties 187–9; escalation of threat 190; and gunboat diplomacy 192–6; and piracy 190–2; and terrorism 191–2; and wartime requirements 187, 189
Peru 239
petroleum: economic use of the sea 31–3, 40, 44–5; gases, transport of 33; and ocean trade 177
Philippine Sea, battle of 14, 26
Philippines 40–1, 239
piracy 190–2
PLARB *see* nuclear-powered ballistic submarine
platforms 96–149; aircraft 138–45; interconnections 149; kinds of 96–8; space 146–9; submarines 132–8; surface warships 98–132; unmanned 145–6
Poland 175, 178, 239
population and foundations of sea power 226
Portugal 239
Posen, B. 206
pre-positioning (PREPO) 210, 214; of military forces 46–7
protection navies 239
Pugh, P. 180–3, 229

Qatar 175, 240

Radar Ocean Surveillance Satellite (Rorsat) 147
Rapid Reinforcement Plan 202–3
Reagan, R. 173, 174, 229
regional force navies 238–9
Rein, Admiral T. 211
remotely-piloted vehicle (RPV) 145
rocket launchers 77–8; against shore 88
role of navies, triangles 233–6
roll-on roll-off (Ro-Ro) 35
Romania 178, 239
Roskill, S. 50
rules of engagement (ROE) 56–7

Saudia Arabia 175, 239
Schofield, Admiral B.B. 49

Index

sea control *see* command of the sea
sea control ships (SCS) 127
sea-launched cruise missile (SLCM) 94
sea power: definition of 3; foundations of 221–32; future of 221–41; and merchant marine 3–4; in the modern world 3–10; role 232–6; typology for navies 236–40
sea routes 38
sea water, properties of 52–3
sea-air-land special forces (SEALS) 118
security, cooperative 154–5
Seicon Patrolling Underwater Robot (SPUR) 146
shipbuilding 178–9; budget for 180–1
shore operations 87–9; and nuclear weapons 93
short take-off/vertical landing (STOVL) 86, 89, 125–7, 128, 141–4, 164
Sidra, Gulf of 100
Singapore 41, 175, 240
'ski-jump' take-off aid 142
small waterplane twin-hull (SWATH) 98, 119–21
socio-political culture and foundations of sea power 226–8
Sokolsky, J. 46
Somalia 240
sonar 79–81
Sound Surveillance System (SOSUS) 78, 80, 146
South Africa 176
South Korea 41, 42, 178, 239
South Yemen 240
space platforms 146–9
Spain 112, 126, 175–6, 238
special warfare craft medium (SWCM) 118
speed of ships 96–7
Sri Lanka 240
stored chemical energy propulsion system (SCEPS) 74
Strategic Arms Limitation Talks (SALT) 67, 154
Strategic Arms Reduction Talks (START) 154
Strategic Defense Initiative (SDI) 87
Strategic Integrated Operations Plan (SIOP) 94
Submarine Advanced Combat System (SubACS) 91
Submarine Command System (SMCS) 91
Submarine Laser Communication Satellite System (SLCSAT) 134
submarine-launched ballistic missile (SLBM) 94–5
submarines 132–8; advantages of 137–8; communications 134–5; and depth charges 77; detection of 79–81; in Iran–Iraq war 133; limitations 132–4; and military power 53–4, 55; and mines 76–7; nuclear, in peacetime 189–90; nuclear powered 135–7; quietness of 135–6; and rocket launchers 77–8; and torpedoes 72–6; warfare 199–201
superpowers: confrontation 154–6; and economic use of the seas 173–4; intervention by 157–9; and 'naval suasion' 166
Supreme Allied Commander Atlantic (SACLANT) 196, 203, 211
Supreme Allied Commander Europe (SACEUR) 94, 211–12
surface effect ship (SES) 118–19
surface-launched air-targeted (SLAT) missile 85
surface warships: aircraft carriers 124–8; amphibious vessels 129–30; battleships 103–4; corvettes 112–13; cruisers 102–3; destroyers 102, 104–9; fast attack craft 99–100; fleet support vessels 130–1; frigates 102, 109–11; fuels 121–3; future form of 114–17, 122; helicopter carriers 128; length-beam ratios 115–16; and manpower 121; minesweepers

99; multi-hull 98, 117;
  nomenclature 100–2; patrol
  vessels 98–9; as platforms
  98–132; second-hand sales of
  111–12; surface skimmers 97,
  117–20, vulnerability 131–2
surface-to-air missile (SAM) 103–4,
  107, 109, 111, 126
surface-to-surface missile (SSM)
  103–4, 108, 110–12, 230
surveillance: in peacetime 196–7;
  satellites 147–8
Sweden 175–6, 239
Sweetman, W. 142
synthetic aperture radar (SAR) 148
Syria 175, 240

Taiwan 111, 178, 239
Tanzania 240
technology, strength of and sea
  power 226–8
terrain contour matching
  (TERCOM) 88
territorial defence navies 239–40
territorial extent and foundations of
  sea power 225
territorialization of oceans 170–1
terrorism 191–2
Thailand 239
Third United Nations Conference on
  the Law of the Sea (UNCLOS III)
  167, 170
threats, escalation of 190
Till, G. 13
token navies 240
Tomahawk cruise missile (TLAM)
  88–90, 93
torpedoes 61, 63–5; anti-submarine
  warfare 72–6
Torrington, Earl of 22
towed offboard active decoy
  (TOAD) 69
trade: and economic use of the sea
  31–8; exports by sea 172–3
Trafalgar, battle of 13
Trinidad & Tobago 175
Tunisia 175, 240
Turkey 239

Turner, S. 12
typology for navies 236–40;
  constabulary navies 240; global
  force navies 237–8; protection
  navies 239; regional force navies
  238–9; territorial defence navies
  239–40; token navies 240

ultra large crude carrier (ULCC) 33
underwater electrical potential
  (UEP) 53
Union of Soviet Socialist Republics:
  Berlin Blockade 57; confrontation
  with United States of America
  154, 156; diplomacy 193–4;
  economic use of the sea 153–4,
  175; fishing 42; foundation of sea
  power 225, 230; as global force
  237–8; Marine Corps 87;
  maritime intervention 158;
  Maritime Strategy for future wars
  208; merchant shipping 38, 40–1;
  missiles and rockets 65–6; as
  naval power 4, 5; nuclear
  weapons 94; space platforms 148;
  submarines 135, 138, 199–200;
  surface vessels 102–3, 110, 112,
  129, 131; torpedoes 64, 75–6
unit cargo vessels 37–8
United States of America: aircraft
  142–3; Berlin Blockade 57;
  confrontation with Union of
  Soviet Socialist Republics 154,
  156; diplomacy 193; economic
  use of the sea 153–4, 173–6;
  foundation of sea power 225,
  226; as global force 237, 240;
  Marine Corps 50–1; maritime
  intervention 158; Maritime
  Strategy for future wars 205–12;
  merchant shipping 4, 40–1;
  missiles and rockets 78; as naval
  power 4, 5; Rapid Reinforcement
  Plan 202–3; shipbuilding 178;
  space platforms 147; submarines
  26, 134; surface vessels 103–5,
  109, 119–21, 125, 128; torpedoes
  73, 76

Index

unitization of cargo 35–7, 40
unmanned platforms 145–6
unmanned underwater vehicle (UUV) 146
Uruguay 240

Veazey, Captain S.E. 120
vehicle carriers 37
Venezuela 175–6, 239
vertical line array DIFAR (VLAD) 80
vertical take-off and landing (VTOL) 102, 142–4
very large crude carrier (VLCC) 33, 216
V/STOL support ship (VSS) 127

warfare: amphibious force 211; carrier battle groups 207–8, 213; convoys 208–9; and economic shipping 203–4; forward operations 210–11; in limited (local) wars 215–17; in Middle East 213–15; and NATO area 209–13; navies in 199–218; nuclear 199–202, 215; in Pacific 213; Rapid Reinforcement Plan 202–3; strategy for 204–6; submarines 199–201; in Third World 217
Warsaw Treaty Organisation (WTO) 57
warships, costs of 181–3
Waters, D.W. 16, 19
weapons systems 61–95; anti-air warfare 81–7; anti-submarine warfare 72–81; anti-surface ship warfare 61–72; command and control 89–92; and nuclear weapons 92–5; shore operations 87–9
Weinberger, C. 160
West Germany 108, 175–6, 227–8, 238
Western European Union (WEU) 161–2, 238
Windass, S. 154
'wing in ground' (WIG) 97
world trading fleet 39–40
World War I, naval strategy 6
World War II: Britain's strategy 50; military power in 49–50; naval strategy 7, 50

Yugoslavia 178, 239

Zumwalt, Admiral E.R. 100

For Product Safety Concerns and Information please contact our EU
representative GPSR@taylorandfrancis.com
Taylor & Francis Verlag GmbH, Kaufingerstraße 24, 80331 München, Germany

www.ingramcontent.com/pod-product-compliance
Lightning Source LLC
Chambersburg PA
CBHW070554300426
44113CB00010B/1251